New Geographies of the American West

New Geographies of the American West

Land Use and the Changing Patterns of Place

William R. Travis

Orton Family Foundation Fellow

Department of Geography

and

Center of the American West

University of Colorado at Boulder

Orton Innovation in Place Series

Washington • Covelo • London

Library of Congress Cataloging-in-Publication Data

Travis, William R. (William Riebsame), 1953–
New geographies of the American West : land use and the changing patterns of place / William R. Travis.
p. cm. — (Orton innovation in place series)
Includes bibliographical references and index.
ISBN-13: 978-1-59726-072-5 (pbk. : alk. paper)
ISBN-10: 1-59726-072-X (pbk. : alk. paper)
ISBN-13: 978-1-59726-071-8 (cloth : alk. paper)
ISBN-10: 1-59726-071-1 (cloth : alk. paper)
1. Urbanization—West (U.S.) 2. Amenity migration—West (U.S.) 3. Land use—West (U.S.) 4. Land use—West (U.S.)—Planning. 5. Regional planning—West (U.S.) 6. Human geography—West (U.S.) I. Title.
HT384.U52W48 2007
307.760978—dc22

2006035636

Printed on recycled, acid-free paper 

Text design by Joyce C. Weston
Manufactured in the United States of America
10 9 8 7 6 5 4 3 2 1

Contents

Foreword

There shall be sang another golden age,
The rise of Empire and the arts
The good and great inspiring epic sage,
The wisest heads and noblest hearts.

Not such as Europe breeds in her decay;
Such as she bred when fresh and young,
When heavenly flame did animate her clay,
By future poets shall be sung.

Westward the course of empire takes its way;
The four first acts already past,
A fifth shall close the drama with the day;
Time's noblest offspring is the last.

—George Berkeley, *Verses on the Prospect of Planting Arts and Sciences in America* (1726)

IT WAS THE Vermonter Frederick Billings who, having left New England in the 1840s to seek his fortune in California as a real estate lawyer, financier, and railroad entrepreneur, purchased 1,200 acres in the East Bay, across from the fast-growing city of San Francisco. With one western town already named after him in Montana, he determined that, rather than call his holding Billingsbroke, as had been suggested by his marketing-savvy peers, he would name it Berkeley, believing that Bishop Berkeley's prophecy of a new golden age would take its firmest root in the American West.[1]

The promise of the New World has always found its boldest expression in the West. Beginning with Thomas Jefferson, whose interest in western settlement as a central means for promoting agrarian republicanism inspired the Louisiana Purchase and the Corps of Discovery expedition, postcolonial Americans, seasoned by the East's thick forests and rocky soils, looked beyond their stone walls and cedar fences and fixed their gaze westward in search of land, economic opportunity, and the realization of the then still emergent American dream. At midcentury, President James K. Polk refashioned Jefferson's western policy into his doctrine of Manifest Destiny, asserting the necessity and rectitude of American expansionism. Later, the historian Frederick Jackson Turner argued in 1893, in his famous "Frontier Thesis," that America's political success—its democratic institutions and civil liberties—depended on the rugged individualism forged by the West's frontier experience.

More recently, the writer Wallace Stegner, in his collection *Where the Bluebird Sings to the Lemonade Springs*, bestowed on the West one of its most celebrated passages when he described the region as "hope's native home, the youngest and freshest of America's regions, magnificently endowed and with the chance to become something unprecedented and unmatched in the world."[2] Although he would at times repudiate the idea of the West as the "geography of hope," believing that certain regional trends, such as water policy, spelled the West's doom, Stegner nevertheless tapped into a deep cultural vein, one that continues to this day to draw all manner of migrants to the region.

This dream helps to explain why the West remains the nation's fastest-growing region. The 1990s and early 2000s saw the latest boom, during which the region's population grew at more than twice the national rate, while its job growth, business starts, and income gains led the nation. The promised land . . . still, all over again.

But at what cost? The West's latest growth surge has caused a new reckoning, not only among westerners but also among others who regard the West as the nation's harbinger. To many who live in or visit the region, the West's signature landscapes and sense of place appear to be receding behind a rising current of development and its predictable effects: a rapidly urbanizing landscape where traffic, sprawling subdivisions, and gentrification are becoming commonplace. According to the American Farmland Trust, 250 acres of Colorado farmland are lost each day to development of one kind or another. The National Resources Inventory reports that across the eleven western states, the percentage of "built-up" nonfederal land increased by half over the past fifteen years. In all, experts believe that roughly one-fifth of the West's private land has been developed for residential, industrial, or commercial uses, and there is no end in sight.

Enter Bill Travis, a western migrant by way of Florida and Massachusetts, who, from his post at the University of Colorado's Center of the American West, has made a career out of analyzing the region's seductiveness, not in literary terms, but in the argot of land use planning and social geography. On meeting Bill Travis, you can't help but feel his passion for the West, as well as his concern. In Travis's eyes, the West is both a singular physical place and a place of mind, a majestic landscape and a precious idea. He believes that today, in the waking hours of the new century, the real West and the West of the imagination have arrived at a crossroads, with land use and development trends putting the region on a path at odds with its enduring values of wide-open spaces, ruggedness, and egalitarianism.

Owing to his unique knowledge of and dedication to the West's landscapes and communities, Travis became the first Orton Family Foundation Fellow in 2005, and *New Geographies of the American West* is the first title in the new Orton Innovation in Place Series to be published by Island Press. The Fellows Program and book series support land use innovators engaged in cutting-edge research, writing, and practice and encourage new approaches to the land use planning challenges facing America's cities, towns, and countryside.

New Geographies lives up to this mandate. As Travis chronicles, the West's sprawling cities and resort areas, fueled by population growth and an expanding economy, are transforming the region's iconic landscapes. "[H]ow much resort growth, suburban sprawl, and rural land subdivision," Travis wonders, "can be accommodated while maintaining the region's remarkable natural wealth—its extensive wildlands and rich biodiversity—as well as its vibrant communities situated in an awe-inspiring landscape?"

Travis is an explorer; the tools of his trade are maps, charts, graphs, aerial photos—anything that will help him understand and interpret the landscape and patterns of human settlement. It is this on- and above-the-ground approach that gives Travis's account its accessibility and clarity. Which is his object. Travis isn't content with simply providing a rigorous study of the West's landscape changes; he seeks to inform people so that they can begin to make better decisions about how, and where, the West will grow.

Building on the insights from his 1997 edited volume, *Atlas of the New West*, in which Travis and his fellow authors described a "new" western economy and culture defined not by mining, logging, and ranching, but instead by a postindustrial, service- and amenities-based economy, Travis seeks to answer this seminal question by defining both the causes and the effects of western land use patterns. With rich analysis and detail, he describes the different kinds of forces creating what he calls the West's four dominant "development geographies": metro-zones, exurbs, resort zones, and the gentrified range. From Denver's sprawl to the ski slopes of Sun Valley to Montana's multimillion-dollar ranches, Travis sees a dynamic, if unstable, patchwork of land use patterns made all the more complex by the region's nagging paradox: the very qualities that continue to attract people and businesses to the West in record numbers are withering under the pressure, like the goose that laid the golden egg.

New Geographies is an academically veiled wake-up call. Acknowledging the age-old failure of American planning institutions and practices to substantially influence western land use patterns, Travis holds up a mirror to the region and asks the (rhetorical) question, "If this is what things looks like, what do we want to do about it?" In his final two chapters, Travis prescribes a diverse set of measures to get the West back on track. From grassroots organizing and advocacy to the use of sophisticated planning technology to rules and regulations, Travis pulls from his holster not a six-gun loaded with silver bullets, but instead an assortment of strategies and tools designed to help strike the proper balance—more like a creative tension—between preserving the West's natural and cultural assets and developing them in support of the region's people, places, and economic possibilities.

The West's greatest truth is that it is a place constantly in the process of becoming, of migrating from where it is to where it wants to be: the "golden age" of Bishop Berkeley's poem, Stegner's "geography of hope." Perhaps there's no end to this migration, this journey, no Pacific coastline marking its terminus, no final act in its epic drama. And isn't that the point? *New Geographies* teaches us that the West's virtues, and its vices, stem from the same source: its open-endedness, its awesome spaces stretching to infinitude, its embrace of new forms, ideas, and lifestyles. What makes the West different, and what will be its salvation, is its willingness to keep forging ahead in search of itself, and in search of a new geography to match.

William Shutkin
President and CEO, Orton Family Foundation

Acknowledgments

Support for this book project was provided through a fellowship from the Orton Family Foundation, with thanks especially to Bill Shutkin and John Fox; the foundation works to improve community development and land use process in the West, in New England, and by example, across the United States. Additional support for my work on western land use change was provided over several years by the William and Flora Hewlett Foundation. The Nature Conservancy first supported the study of ranch ownership change described in chapter 7, and Mike Clark, now with Trout Unlimited, further supported that effort. My thanks go to Patty Limerick and the crew at the Center of the American West for providing a supportive nexus for research and practice on western issues; Charles Wilkinson and Ed Barber for their ideas and encouragement; and Hannah Gosnell, Julia Haggerty, and Jessica Lage for their many contributions to the study of western lands and collaboration on the research described in chapter 7. I have learned a lot from them. My appreciation also goes to Dave Theobald for his great insights into the geography of land use; Tom Dickinson at CU's Institute for Behavioral Science for his technical skills, from GIS to cartography applied to many of the figures, and his love of western landscapes; Mark Haggerty for mixing fly fishing with conversations about development in the west; Geneva Mixon for data analysis, graphics, and permissions; Luther Probst and Myles Rademan for insightful comments on drafts; and Nancy Thorwardson and Molly Holmberg for help with maps and illustrations. At Island Press, many thanks to Jeff Hardwick, Shannon O'Neill, Jessica Heise, and especially Heather Boyer, who brought this book to fruit and made it better. Finally, thanks to my mother, Gloria Riebsame, tireless chauffer and supporter of my early scientific and academic interests, procurer of student loans, and later in life, fellow explorer of western landscapes.

New Geographies of the American West

Introduction

Building a Better Mountain?

A DEVELOPMENT BOOM washed over the American West in the 1990s and early 2000s.[1] The results were especially conspicuous in the region's ski resorts, all of which seemed to be adding ski terrain, lifts, and expanded base villages simultaneously in the largest region-wide expansion in decades. Entire new base villages went in at Mammoth Mountain and Kirkwood, California, and at Copper Mountain and Winter Park, Colorado, and new lodges, lifts, and terrain were added at Steamboat, Park City, Beaver Creek, Vail, Heavenly, Big Mountain, Mount Crested Butte, Snowbird, Solitude, White Pass (Washington), Williams (Arizona), Mount Ashland (Oregon), and Angel Fire (New Mexico).[2] Improved air service came to Big Mountain, the Tahoe resorts, Jackson Hole, and Sun Valley.[3]

The ski industry's effort to build better mountains was only one manifestation of an all-encompassing wave of development. No landscape was left untouched as new geographies of development spread across the region. Driving through the West, you can easily see the transformation, starting from the new lofts in downtown Denver and Boise, past new suburbs that are now whole cities in their own right, out to the small towns newly "discovered" by urban refugees, on to the West's rangelands bedecked with trophy ranches, and finally to the region's wildlands, now fringed by a settlement ring of rural mansions and ranchettes.

In the long view, this boom was only one of many peaks in the enduring expansion of western land development. In a sense, the West has boomed for more than a century. It has grown in population faster than the nation as a whole in all but one of the last ten decades. This growth has been spurred by demand for resources such as gold and silver, oil and gas, coal, timber, and cattle, and more recently by the desire to live in the nation's fastest-growing cities, towns, and rural areas, all set in a dramatic landscape of mountains, deserts, and canyons. The West's rapid development raises new debates about land use: how much resort growth, suburban sprawl, and rural land subdivision can be accommodated while maintaining the region's remarkable natural wealth—its extensive wildlands and rich biodiversity—as well as its vibrant communities situated in an awe-inspiring landscape?

This book takes up those questions by diagnosing land use trends and taking stock of changing landscapes and communities. It also proposes ways in which future development can sustain the West's ecological and cultural values. In an examination that is part regional geography, part land use analysis, and part planning prescription, I lay out the development patterns that are changing the face of the West and appraise their underlying driving and shaping forces. From edge-city office parks to ritzy ski resorts pressed against the wilderness, swelling land development threatens the ecological integrity of the region and alters the social functioning of its human communities. The region is far from "built out"—it remains rich in open spaces and natural areas—but development is also more pervasive than it may at first appear, the effects of growth reaching out to transform even the natural landscapes and the processes that shape them, such as wildfires and forest succession.

The development patterns analyzed here presage trends well into the twenty-first century and raise the question: can we put western development on a trajectory more appropriate to the region's landscape values? Traditional land use planning has done little to mitigate the negative effects of rapid western growth; indeed, planning in the West is mostly about encouraging and enabling growth and land development. Yet concern over growth is part of daily conversation among westerners. Some western states have entertained constitutional amendments meant to slow growth, and others have passed legislation mandating "smart growth." Letters to the editors of newspapers from San Diego to Helena speak of (and often grieve over) lost views, crowds where once there was solitude, skyrocketing house prices, and farms and ranches subdivided. The heart and soul of the West is being whittled away by new suburbs, new ski resorts, and new ranchettes.

The Changing West

The American West—especially the roughly 1 million square miles of mostly dry, rugged terrain from the front ranges of the Rocky Mountains westward to the Sierra Nevada and Cascade ranges—remains, to many Americans, the land of wide-open spaces, cowboys and miners, and national parks. Even its cities, such as Tucson, Salt Lake City, Denver, and Boise, although in many ways similar to cities everywhere, are seen as slightly exotic outposts on the "frontier."[4] For the past two centuries, the West has been the focus of the nation's search for natural resources, with its wealth of silver, gold, grazing, timber, oil, coal, uranium, and natural gas, as well as providing the isolation that makes building and testing nuclear bombs (and eventually storing the waste from nuclear power plants) a bit safer. It is also where Americans have created the nation's largest national parks, national monuments, and wilderness areas.

For some time now, most of the development occurring in the West, both in the interior and along the Pacific coast, has had little to do with natural resource extraction. Mining, logging, and ranching are now relatively minor parts of the western economy. An economically diverse postindustrial regime of services, information technology, light manufacturing, tourism, and retirement now drives growth. These economic changes have also transformed the region's land use patterns and have altered its long-standing land use battles. The suburbanization indigenous to the Midwest and South now spreads across its foothills and canyons, and arguments over urban sprawl, affordable housing, ski area expansion, water transfers from farms to cities, and residential development in wildlife habitat and fire-prone forests drown out debates about clear-cutting, strip mining, and livestock grazing.

The emergence of this "New West" is difficult to date. Some historians mark the region's postindustrial epoch as beginning with the New Deal in the 1930s, further spurred by World War II.[5] But the region's modern identity came most distinctly into sight during the 1960s and 1970s. Law professor Charles Wilkinson tied his evocation of the New West to the opening of Vail, the nation's biggest and richest ski resort, in 1962 and to the bitter fight over Colorado's plan to host the 1976 Winter Olympics: "The creation of Vail, the rise of a high-stakes recreation industry, and the dispute over the Olympics epitomized a new dynamic in the region." He writes of America's "deepened passion for mountain terrain coupled with its newfound love of deserts and plains."[6] Americans want to live in and preserve these landscapes simultaneously, creating new tensions that have erupted into new battles: over water supply and Mono Lake in California, land use and urban sprawl in Oregon, and expanded ski areas in Colorado.

Nothing seemed to express the region's new, and newly problematic, economic dynamics better than Las Vegas, the nation's fastest-growing city at the millennium, which author Mike Davis labeled a "hyperbolic Los Angeles—the Land of Sunshine on fast-forward" and the advance guard of "an environmentally and socially bankrupt system of human settlement," a sprawling, water-greedy "apocalyptic urbanism in the Southwest."[7]

The Footprint of Development

Apocalyptic? Well, maybe. Western land development is certainly sprawling and water-greedy, although Las Vegas itself is one of the densest cities in the country. Spreading residential and commercial land uses are transforming the West's emblematic landscapes: its mountain fronts, its great swaths of rangeland, and its desert canyons. At risk is wildlife habitat, biodiversity, nurturing human communities, and the sense of place that comes from the West's terrain, climate, and history.

This landscape transformation is at a critical juncture. As the fastest-growing region in the United States, the West is at risk of losing the qualities that make it unique. Indeed, exigencies of climate, geology, and geography make modern western development especially harmful to ecological health. The most ecologically valuable land is especially attractive to development. Preserved public lands, such as national parks, draw residences and businesses to their fringes and feed a growing recreation and tourism economy that further invades the wilderness.[8]

In the postindustrial economy, people and businesses find increased freedom to choose their locations for "quality of life," but these newly desirable locations are often in the foothills, part of the "wildland-urban interface" in the lower forest boundary, and are routinely swept by wildfires. Run-ins between people and wildlife, and between houses and wildfires, escalate.

The clear message from the fields of biology and landscape ecology is that habitat patterns—not just the amount of habitat—determine biodiversity and ecological health. Similarly, the spatial patterns of development can sustain or weaken the social web of connections that creates functional and nurturing communities. Furthermore, development's "ecological footprint" is larger than the area physically covered by houses, parking lots, office buildings, malls, and gravel pits. Western development requires more water than local watersheds can provide, so communities reach into distant river basins for water. Development demands energy, much of it extracted from the public lands, and it stretches roads like a net over the region. Relatively natural areas near developments are less inviting to wildlife, more subject to invasive species. The enlarging human imprint of regional development pervades even remote wildlands, where, for example, wildfires cannot be left to burn because they might eventually threaten the subdivisions that have crept up to the boundaries of the public lands.

Recent attempts to estimate the "footprint" of development suggest that the land affected by a North American city is somewhere between ten and twenty times the actual built-up area.[9] But we cannot simply multiply the area of western cities by ten or twenty to get their total "footprint" on the region; no regional standard exists for such calculations. Moreover, the effects of development in the West depend on geography. Some areas given over to intense recreation, waste disposal, or highways, for examples, are less sensitive than others, less critical to regional ecological integrity and social well-being. Although determining this sensitivity—choosing, as it were, the best land to develop—is an audacious act requiring more than a little wisdom, it is a crucial part of any realistic prescription for western development.

Development and the Heart and Soul of the West

Rapid development is also changing western society. Sprawl, lack of affordable housing, and traffic jams are changing the mentality of the West. Westerners, even relative newcomers, lament the landscape and social values squandered as roads are carved into mountainsides and ranches subdivided. Many are working to protect what's left of the region's natural wealth and working landscapes, but weak planning institutions and a resurgent property rights movement frustrate their efforts. Growth's critics argue, with some justification, that rather than ensuring quality of life, government in the West mostly promotes further development with pro-growth programs of all sorts, from tax breaks to water projects. Much of what we lament about modern development in the West was planned. Antigrowth, slow-growth, and even "smart growth" forces are weak, their campaigns outmaneuvered by local and regional growth machines.

The critics of land use planning, antisprawl campaigns, and open space programs argue that growth management limits choice, tax base, and the market's ability to provide a range of housing. They instill the fear that any effort to slow growth will lead to an economic bust, that jobs will disappear from western communities. They would have us believe that our choice is sprawl or stagnation. But several studies suggest that their pro-sprawl case is terribly flawed. Communities that plan carefully, that preserve open space and the ecological services of nearby natural areas, are more economically and socially successful. Those communities attract businesses and jobs and have the resources to invest in community well-being. Instead of subsidizing sprawl with infrastructure and other forms of public investment, why not invest public resources in smart growth, housing affordability, and open space?

Time for Change

There is time for change. Despite the rapid spread of western development, the mountains, basins, deserts, and canyonlands of the West are still by and large open country, sparsely settled, for the most part, outside its few large cities (such as Phoenix, Salt Lake City, Las Vegas, and Denver) and medium-sized towns (such as Reno, Boise, Grand

Junction, Bozeman, Helena, Colorado Springs, and Tucson). Nighttime satellite images reveal that, from the Rocky Mountains west to the Sierra Nevada and Cascade ranges, the explicit footprint of development is still small compared with that in other American regions. Open land is the region's chief asset, providing habitat for something close to the full suite of its natural biological heritage as well as a compelling matrix for its human residents.

There is ample opportunity for change. Because most communities in the region are growing, their planning boards and elected officials are constantly reviewing development applications. Indeed, from Greeley, Colorado (the fastest-growing city in the United States during 2000–2003) to Barstow, California, development applications are stacked up like airplanes at a busy airport. This is precisely the situation in which change can occur, in which local and state officials can implement better development practices, set aside open space, and make development pay not only for the direct services it requires, such as roads and sewers and schools, but also for the indirect services of affordable housing, open space, and recreational lands.

Many western communities are surrounded by farmland and ranchland that provides open space, character, and even the potential for local food supplies. Others lie in a geography of public lands. Although they are not subject to private development, the national forests, national parks, and wildlife refuges do host timber cutting, mining, grazing, roads, and recreational developments that can and do degrade the land. Such uses must be curbed as private lands are developed and the public lands become the main reserve of habitat on which the region's ecological well-being relies.

Much more development is on tap: the region's population will double in the next forty to fifty years, and up to half of the remaining developable land is on the chopping block, slated for houses (at low to high densities), offices, warehouses, ski villages, golf courses, shopping malls, highways, airports, and the other accoutrements of modern American development.

The stage is thus set for a struggle over the future of regional landscapes, a struggle that pits open spaces valued for social and ecological reasons against growth and development. This struggle is not new in the history of American land use,[10] but the battle lines are especially sharply drawn in the West, and the outcome there will determine to a great extent the ecological health of the nation and the social health of a region.

If the current trajectory of western development undermines the region's natural and social sustainability, then how should it be altered? I worry about rapid development, especially in my own western place, on Colorado's Front Range. I am concerned that we will be sorry after we have, in historian Patricia Nelson Limerick's phrase, "fully deployed conventional American culture in [this] unconventional landscape."[11] The region is rapidly becoming just like the rest of the country, with sprawling suburban cities, the standard malls and office parks, and cookie-cutter housing developments on subdivided farms and ranches. Residents respond to the region's booming development with attitudes ranging from resentment to grief. Their calls for better land use planning, and even limits on growth, have had limited effects thus far, but their efforts show that westerners want something different.

There is still time to alter the settlement trajectory of the West, but the effort must be strengthened quickly because much more development is coming to the region. This argument rests in part on an interpretation of the region's history that challenges the conventional view: the western development experience is not a series of boom-and-bust cycles, but rather a history of cumulative expansion that will probably continue for the foreseeable future. The shape of that expansion is not predestined, however. It will be shaped by people, by westerners who want to live in communities that nurture social life but also let in a little wildness, who want to break the vicious cycle of development and wildfire on the urban fringe, and who are willing to take on the responsibility of settling a region without pauperizing its natural capital.

~ ~ ~

Looking out from my suburban vantage point on the edge of Boulder's protected open space, I find it more difficult than many other western writers to criticize settlement patterns or to prescribe how westerners should live. Still, the new geographies of the West, driven by a complex set of personal preferences, economic logic, and government incentives and disincentives, bring obvious problems to a region enriched with widely valued ecological and cultural landscapes, landscapes marked preeminently by an openness and naturalness, a western-ness that embraces limited human settlement. Ultimately, this stock-taking does question whether we are, in any sense, creating a better place as we transform the American West, and it seeks to persuade the reader to reject orthodox development and planning doctrine and to imagine and enact an alternative regional vision.

Part One

Understanding Growth, Development, and the Changing American West

The first step toward prescribing better development patterns for the American West is a clearer diagnosis and prognosis, a better sense of where we're at and where we're going. Key to this analysis is an understanding of the region's historical development trajectory and of the forces driving and shaping its current land use patterns. Population growth, burgeoning wealth, mobility, and changes in construction technology all drive and enable the sprawl of development across the western landscape. The region has moved beyond its history of "boom-and-bust" economics, and there is little reason to expect anything other than a rapidly spreading development footprint in a region seen as an attractive place to live and do business.

Figure 1.1 Places in the American West.

1 The Long Boom of Western Development

THE AMERICAN WEST is in transition. Population growth at more than twice the national rate and leading rates of per capita job growth, business starts, and income growth attracted national attention to the region in the 1990s.[1] The national media offered a steady stream of stories hyping the West's latest development boom while also expressing residents' concerns about urban sprawl, displaced wildlife, and stretched water supplies.[2] Those concerns flooded onto the editorial pages of the major newspapers, into local public meetings, and onto election ballots. Predictably, the growth surge also evoked allusions to the West's frontier history. Some observers saw this latest development rush as another round of the region's infamous boom-and-bust cycle; they argued that the current development rush would bust, just like the gold rushes, cattle booms, and energy bull markets before it.[3] The implication: we had better not do anything to squash the boom and thereby risk making the inevitable bust even worse. Growth boosters spent millions of dollars instilling this fear in voters during campaigns over growth management initiatives on the 2000 ballot in Arizona and Colorado.

The boom did slow, briefly, along with the national economy, in 2001–2002, giving many political leaders an excuse to ignore public concerns about sprawl, traffic, and loss of open space. But most western places (fig. 1.1) continued growing rapidly right through the national slowdown (especially large cities such as Phoenix and Las Vegas and charismatic rural areas such as those near Yellowstone

National Park), and the region continued to outpace the nation in growth. This sustained growth belies the expectation, based on a misreading of western history, that busts inevitably erase the effects of growth, that regional economic expansions always end in contractions. This view bears closer scrutiny as we contemplate decades of future western development.

I use the words "boom" and "bust" here with some hesitation because in this chapter I will argue that we have mischaracterized western development history as a repetition of booms and busts, as a cycle of growth followed by retrenchment. This misconception is especially relevant to land development: except in very isolated cases, land development does not come and go (ghost towns are sufficiently rare to be tourist attractions). Instead, development subjects land to increasingly intense uses that permanently transform the natural and cultural landscapes, even after growth spurts end. The West's geography is permanently inscribed more by boom than by bust.

Episodes of rapid population growth and land development—the "booms" of western history—do indeed subside, and perhaps a few places actually lose some population in local downturns. But the trajectory of western development is much more cumulative than the cyclic historical model implies. Even in the last episode that westerners called a "bust"—the end, in the early 1980s, of the big run-up in energy development that was spurred by high oil prices—no western state showed a permanent loss of population (that is, no state had fewer residents in 1990 than it had in 1980), and many areas still outpaced national growth rates. It was not long, moreover, before the word "boom" was again heard in the West, in reference to the 1990s development rush, which did not slow until after the new millennium.

I started to study western development closely during the 1990s boom, and I began writing this book during the slowdown in regional development brought on by the national economic downturn that marked the second and third years of the new millennium. Spanning these different patterns of growth strengthens this analysis by allowing a more measured assessment of enduring regional land use problems, not just the effects of a particular boom decade. Yet I must report at the outset that, even in a period of slower growth, I was still impressed with the region's potential for future rapid development and its ability to grow even when the national economy languishes. As of this writing, in 2006, the Interior West's population is still growing faster than that of any other region of the nation, and the fastest-growing places in the nation are immediately east of me in Weld County, Colorado.

If anything, the slowdown between 2001 and 2002 and the subsequent return to faster growth showed that the region's development trajectory remains fundamentally unaltered. It did provide a breather, offering westerners a chance to assess the landscape changes visible across the region, to catch up with growth, and maybe even to do some effective planning and growth management. But ratcheting growth, well illustrated by the increase in the region's share of total U.S. population over time (fig. 1.2), is the overarching

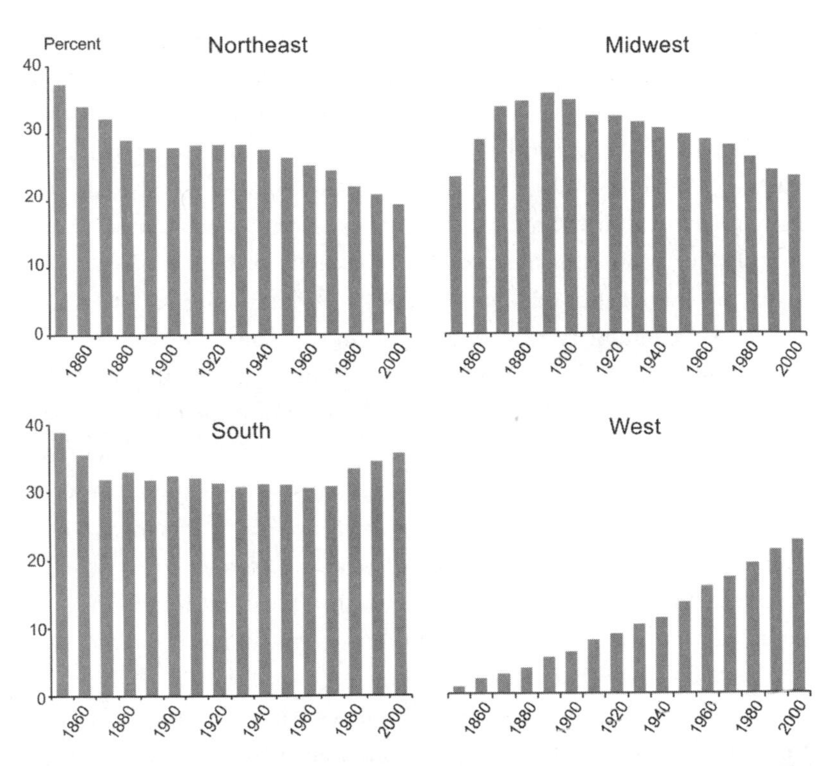

Figure 1.2 Share of total U.S. population by census regions, 1850–2000. The West stands out as having grown faster than the other U.S. regions through most of its settlement history. *(Modified and updated from George Masnick, "America's Shifting Population: Understanding Migration Patterns across the West,"* The Rocky Mountain West's Changing Landscape *2, no. 2, Winter/Spring 2001: 8–14.)*

fact of western development and its centuries-long, cumulative spread across the landscape.

The Start of the Long Boom

Since humans arrived some 13,000–15,000 years ago, most of the West has been occupied, at least seasonally, at increasing human densities, resulting in a spreading transformation of its landscapes. Some early, intense developments did come and go. The citylike settlements of the Chacoans and Hohokam in the Southwest fell into disrepair after 1200 A.D. These ancestral Puebloans didn't "disappear," although they did de-urbanize for the most part. Some were absorbed into a more dispersed Southwestern culture; others maintain settled pueblos along the Rio Grande to this day.

The European invasion, starting in the 1500s, set in motion the development trajectory that marks the region today. The rate of historical development has varied, quickening and slowing, in episodes that came to be seen as the endemic western pattern of "boom and bust." Episodic economic and demographic booms certainly did put new people and investment into western places faster than prevailing frontier diffusion rates. The Spanish incursion up the Rio Grande Valley in the early 1500s brought on rapidly expanding irrigated agriculture and settlement. After gold was found on the American River at Sutter's Mill on January 24, 1848, the European American population of California increased from roughly 15,000 to over 60,000 in just one year, and reached some 220,000 by 1852.[4] A few years later, gold was found in the southern Rocky Mountains, and by the summer of 1860 over 5,000 miners a week, some originally on their way to the California diggings, inundated the Rockies.[5] Next it was silver in Montana and Idaho. Although many of the western towns that got their start in mining never became big—and today a few are even abandoned—the wealth that flowed in and out of the gold, silver, and copper fields spurred the growth of cities such as Denver, Boise, and San Francisco, cities that then fueled further growth. Moreover, mining required railroads and wagon roads, and this infrastructure also began to serve ranching and logging (even as the minerals played out), which were operating pretty much wherever the requisite resources existed by 1900.[6]

Still, the economic regime of gold, cattle, and timber—like most economies based on raw materials—was inherently unstable. Although a bust inevitably seemed to follow each boom, precipitated by a crash of commodity prices, a change in consumer tastes, or a catastrophic event (as when the open-range livestock industry virtually ended in the drought- and blizzard-induced "big die-up" of 1886–1887),[7] these setbacks were not permanent. Many of the forty-niners left the California gold fields as the placer deposits played out, but they didn't leave California, and the region never returned to anywhere near its pre-rush population. There were other reasons for settling in California: farming, trade with Asia, and the strategic occupation of territory wrested from Mexico. The underlying settlement regime was more stable, and more cumulative, than the region's image as boom-and-bust territory suggested.

Commodities to be shipped out were not the West's only attraction; land itself lured people, especially during the national demobilization following the Civil War. Historian William Robbins described the West as looming large in the nation's post–Civil War economic outlook. It was seen "as an investment arena for surplus capital, as a source of raw materials, and as a vast vacant lot to enter and occupy."[8] To settle the nation's western reaches quickly and to disperse demobilized armies, the federal government encouraged western settlement by surveying the region and offering homesteads for a small fee. It also granted land for 25 miles on either side of the rails to the transcontinental railroad companies, which then further encouraged settlers to fill the space and, not incidentally, their freight and passenger cars.[9] If early settlement faltered due to climate or economics, the government stepped in to shore it up, with outright subsidies and with infrastructure such as dams, irrigation projects, and crop insurance.

Western land speculation was self-reinforcing. Although the history books focus mostly on tangible resources such as cattle, timber, and gold, the deep-seated expectation that land values would increase was, and still is, a key driving force in western development. Historian Patricia Nelson Limerick put it this way:

> If Hollywood wanted to capture the emotional center of Western history, its movies would be about real estate. John Wayne would

> have been neither a gunfighter nor a sheriff, but a surveyor, speculator, or claims lawyer.[10]

He might also have been a real estate broker. The nation's "manifest destiny" was rooted in acquiring, dividing, and reselling property, and settlers in the West took to it with a passion and rapidity that kept them ahead of the government's efforts to survey land and record land claims. Although we tend to see the "land rush" as a uniquely frontier phenomenon, there is little difference between the speculation-driven development in the homesteading era and the late twentieth-century run-up in western resort real estate, or the more mundane appreciation of suburban homes. Speculation feeds, and feeds on, the rational expectations (and, occasionally, the irrational exuberance) of landowners, who then push for, or at least tolerate, the pro-growth postures of local leaders because their policies sustain land appreciation.

Homesteaders brought new land uses and technologies that transformed the region's ecology and economy in more profound and enduring ways than their predecessors had, imposing a land regime based on private property parceled out and deeded for permanent occupation and development.[11] With government support, they introduced domestic livestock grazing essentially everywhere (degrading and transforming the vegetated landscape), cut most old-growth forests, and diverted water from streams onto fields and off to the cities.

Conjoined national purpose and individual ambition yielded an impressive result: the progressive settlement and industrial and mercantile development of a region that explorers and surveyors had pronounced, in the early 1800s, essentially uninhabitable.[12] The federal government maintained much of the settlement and development momentum by building reservoirs, constructing roads, and encouraging extraction of resources—from grass to silver—on public lands at below-market fees.[13] Even its withdrawal of some lands from homesteading to be maintained as forest reserves, national parks, and common grazing allotments spurred development by reducing the supply of private land and imparting special values to some lands.

In a sense, a great federally boosted development project, not unlike the grandiose five-year "new lands" development plans of the

Soviet Union (except that it was more effective), was deployed in the American West. Government support, in the past and today, is one reason why the West's settlement history is more cumulative than boom-and-bust, with more growth than recession. Even the great national economic bust of the 1930s occasioned something of a development boom in the West: while the Great Plains was losing population, the Interior West and Pacific states grew. The Relief and Construction Act of 1932 and other make-work elements of the New Deal brought a flood of the unemployed from eastern and midwestern cities to the West. The region outpaced all others in its per capita share of federal funds for construction of irrigation projects, roads, public buildings, trails, and even tourist facilities.[14] Four of the biggest dams in the world today were under construction simultaneously in the West in 1936, all with federal money, equipment, and planning. Farmers driven out of the Great Plains by drought moved to California. Unemployed easterners joined the Civilian Conservation Corps (CCC). My father, Frank Riebsame, was shipped from Philadelphia to a CCC camp near Flagstaff, Arizona, where he fought fires, built trails in the Grand Canyon, and generally had the "time of his life" in the far West.

The West continued to grow when World War II ended the Depression. Both hot and cold wars spurred even more federal investment and set the stage for the postindustrial development described in the rest of this book. Historian Richard White assessed the region's attractiveness during World War II as follows: "All the old liabilities of the West suddenly became virtues. Vast distances, low population density, and arid climate . . . became major assets as military planners scrambled to locate new military bases."[15] Wartime and postwar policies further opened the region, providing infrastructure, people, and industry. Geneva Steel Works was moved to Utah, away from a Pacific coast vulnerable to attack. The atomic bomb was designed, built, and detonated at secret sites in the interior. The Cold War kept the nuclear and strategic mineral and manufacturing sectors alive: rocket engines built and tested on the shores of the Great Salt Lake would carry atomic warheads built in secret complexes on the Columbia River and atop New Mexican mesas, then tested in the Nevada deserts. Many military retirees and defense workers decided to stay on, and the country's abiding westward migration quickened, compelled by jobs, but also by climate and scenery.[16]

The Last Great Boom and Bust?

Although I paint a picture here of a region experiencing much more boom than bust, a recent episode of western growth is often cited as revealing the continued grip of the archetypal boom-followed-by-bust pattern.[17] The 1970s oil embargo by the Organization of the Petroleum Exporting Countries (OPEC) raised global energy prices dramatically and pushed American policymakers to emphasize domestic energy exploration and production. In what Charles Wilkinson called the "Big Build Up," oil riggers, geologists, and coal miners inundated the West—where most of the untapped energy in the contiguous United States was located—from the coalfields on the Great Plains to the oil wells off the California shore.[18] Western population grew by almost 40 percent in the 1970s.[19] On the Colorado Plateau that Wilkinson writes about, the rush included the construction of eleven coal-fired plants generating electricity to meet the needs of Phoenix, Las Vegas, Los Angeles, and Salt Lake City. The twelfth proposed plant, on the Kaiparowits Plateau, finally generated enough public outcry—against the pollution, water demands, and landscape effects of mining, all for producing electricity to be exported off the plateau—that California Edison cancelled the project.[20]

When the oil cartel could no longer sustain high prices in the face of mounting global supplies (evoked by OPEC's artificially high price), the cost of energy plummeted, and the West's energy boom busted in the early 1980s. Boomtowns such as Gillette and Rock Springs, Wyoming, and Rifle and Rangely, Colorado, descended into the development doldrums, their new roads and sidewalks—laid out to accommodate subdivisions for the energy workers and their families—leading to sagebrush slopes instead of houses. The breakneck development, with all the accompanying grief over inadequate housing and services, negative effects on the environment, and rising crime, suddenly seemed desirable in the post-boom hangover of lost payrolls, home loan defaults, and out-migration.[21]

But even this bust—and the term is at least locally accurate in this case, especially in western Colorado, where huge federal subsidies had been sunk into an ill-fated scheme to bake petroleum out of oil shale—did not depopulate the energy boomtowns by even a quarter. Many parts of the West, such as the amenity-rich zones around national parks and the urban swaths from Southern California to the Colorado Front Range, actually *gained population* right through the bust. All eleven western states, even those that bore the brunt of the oil price collapse, ended the decade of the 1980s with more people (and less open space) than they had at its start. While most rural counties in Wyoming and Montana lost population in the 1980s, the counties around Yellowstone National Park, home to attractive towns such as Jackson and Bozeman, still grew at twice the national rate, as did counties with larger towns and cities. The region's largest cities also felt the loss of energy jobs, especially those in which the energy firms had planted administrative offices. Although much was made of empty office buildings in Denver, and although the city's population *declined* by roughly 10,000 during the 1980s, the larger, mostly suburban, Denver metro area still *grew* by over 400,000 residents in the same period. Downtown Denver had more than made up its lost population by 1993.

Some western observers see the 1970s–1980s energy boom and bust as the last great commodity cycle in the West—as the end, if you will, of western cyclic history. From a regional perspective, however, the bust seemed strangely ineffective. Even its central foundation, energy extraction, did not actually bust. Western coal production reached an all-time high at the start of the new millennium (growing from 630 million tons at the height of the energy boom in 1978 to 1.1 billion tons in 2001). The western states now produce over half of the nation's total coal output.[22] Natural gas extraction grew right through the 1980s and 1990s, from 20.9 trillion cubic feet (tcf) at the height of the energy boom in 1976 to 23.7 tcf in 1999; this fuel became even more sought after in the 2000s by a world increasingly worried about global warming. Aztec, New Mexico, in the heart of the San Juan Basin gas fields, has for almost two decades resembled the bustling oil towns of Wyoming's Overthrust Belt during the OPEC-induced boom of the late 1970s, which hardly lasted five years. Drilling rigs light up the horizon, service trucks prowl the dirt roads from one wellhead to another, and the town's businesses, from car dealerships to restaurants, thrive. Wyoming, the slowest-growing western state in the 1990s, beat the job creation rates of its neighbors in 2000–2002 due in large measure to increased natural gas production.[23] Only oil, which reached its maximum U.S. production in 1970, flowed less effusively in 2000 than it did in the 1970s. There may have been an energy "bust," in the lingo of popular western history, and at this writing there are the makings of another energy boom, but it is difficult to discern such swings in regional economic and demographic indicators. Western energy extraction simply seems to grow through "boom" and "bust" alike.

Even with energy being produced at record rates, other drivers of regional growth have overshadowed it. The quieter rise of both professional and low-skilled service jobs, tourism, and the region's high-tech economy set the geographic stage for even more substantial western population growth in the 1990s and beyond.

The "Amenity Gold Rush"

For some time now, the West's resource extraction economy has been eclipsed by the growth of a long list of commercial and professional services, information technology, telecommunications, recreation, retirement, and its oldest industry: land speculation. By the mid-1990s, only 19 counties out of the 400 in the West could claim that at least one-third of their jobs were in mining, manufacturing, logging, farming, and ranching.[24]

The media, and some economic gurus, have argued that multiple sectors of the postindustrial economy, from high-tech to recreation to retirement, have been nurtured by the region's geographic smorgasbord of charismatic and amenity-rich landscapes, its dynamic towns and cities, and its pastoral, ranching, fly-fishing rural areas. This is the New West that Americans find an increasingly attractive place to live.[25]

A Regional Makeover

This economic transformation, still under way, can be explained as the inevitable outcome of globalization and the shift to a service-based economy in the industrialized countries. Some of its power comes from the decline of prices for commodities in a globalized economy, in which countries with cheaper labor and resources can compete with western farmers, ranchers, loggers, and miners. Western land is losing value as a vessel for resource extraction.

Economist Thomas Michael Power drew a portrait of this economic transformation in his book *Lost Landscapes and Failed Economies: The Search for a Value of Place*. He also highlighted tensions between the West's declining extractive industries and its new economic geography:

> Empirical analysis shows that mining, timber, and agriculture make a much more modest contribution to local economies than is usually assumed. The ongoing transformation of local economies, including technological and market changes, has drastically reduced the relative importance of such industries. As a result, rather than being a source of economic vitality they are likely to play a declining and destabilizing role in local economies in the future.[26]

Power found that local and state leaders often failed to recognize the economic transition and that their adherence to old economic models actually hurt communities.

At several meetings in western communities, I watched Power offer his analysis and his message—that in order to benefit from the West's new economy, communities had to protect rather than exploit their natural landscapes—and saw the disbelief (and sometimes resentment) with which his ideas were often met. In addition to the farmers, ranchers, loggers, and miners who couldn't or wouldn't accept Power's argument, many western suburbanites, whose livelihoods were based in the modern economy and who enjoyed the landscape amenities he extolled, had trouble recognizing that the cowboys and miners were not the backbone of the western economy. It was as if they saw their urban lives and occupations as somehow less real, less vital to regional development, than those of the "people of the land." Perhaps they also believed the other arguments that critics often throw at Power: that the old economy paid better, that the new economy was only temporary, and that when the price of oil, gold, timber, and red meat finally "recovered," the West would once again prosper from its natural resources.

In response to these criticisms, Power co-wrote, with economist Richard Barrett, *Post-Cowboy Economics: Pay and Prosperity in the New American West*. In it, Power and Barrett strengthened the argument with a hard-nosed analysis of the numbers that proved the worth of the new economy and the harm of propping up the old. They took on the income debates and the fears of boom and bust that seem hardwired into the western psyche. Besides simply pointing out that personal income is growing faster in the West than elsewhere, they argued that people have long counted quality of life in their economic decision making, and that the West's attractiveness is a key asset to figure into any accounting of its economy. They concluded, in a reference to the West's cut-and-run past, that "the region's growth is sustained by its attractiveness rather than by glittering opportunities to strike it rich."[27]

Although this economic transition is widespread and inevitable, many local development directors, along with boosters of extractive industries such as ranching, have been slow to incorporate it into their thinking. Their reluctance results in a salient tension between the Old West and the New West, which a political culture stuck on the real and imagined benefits of an extractive economy abets. State legislatures fight progressive reforms, of everything from wildlife management to water allocations, that they think will hurt ranching, logging, and other resource industries. In their calculus, water for recreation simply does not have the same importance as water for crops.[28] Many local leaders feel obliged to prop up their copper smelter or timber mill, if they still have one, taking an approach that Power calls seeing the economy "through the rearview mirror."[29] He argues persuasively that attempts to shore up the extractive economy are doomed by global economic trends (raw commodities are in oversupply most of the time) and actually hurt local communities' chances to thrive on the environmental resources that, properly conserved, attract new businesses, entrepreneurs, and a well-educated labor force.

Unsettling the Old West By the turn of the millennium, the Interior West at last seemed to be catching up with the West and East coasts, transforming itself from a natural resource colony into something approaching a mature postindustrial regional economy. The West is so attractive and well suited to the structure of the postextractive era that it could someday come to dominate other U.S. regions. *Time* magazine hyped this notion in a 1993 cover story, "Boom Time in the Rockies." The cover illustration was of the Maroon Bells (mountain peaks near Aspen, Colorado), with their reflection on a mountain lake (with two hikers looking on) transfigured into Denver office towers.[30] The message was obvious: the West offers a unique blend of wilderness, recreation, urbanity, and business opportunity.

Time's story, which described families escaping urban blight to watch sunsets behind snowcapped mountains, took a stab at defining the raison d'être of the region's new development: "The Rockies' new ethos manages to combine the yearning for a simpler, rooted, front-porch way of life with the urban-bred, high-tech worldliness of computers and modems."[31] As indicators of the region's florescence, *Time* cited Denver's new mega-airport and high-tech office parks, Salt Lake City's information economy and its hosting of the 2002 Winter Olympics, Montana's growing service economy, and Jackson Hole's housing shortage. It compared these trends to the mid-1980s energy bust that emptied some of Denver's high-rises and briefly made real estate affordable in Crested Butte, Bozeman, Ridgeway, and other mountain towns now too expensive even for middle-class home buyers.

But this was not just a Rocky Mountain high; much the same could have been written (and eventually was) about Reno, Phoenix, or the burgeoning corridor from Seattle to Bellingham. Indeed, a few years later, *Newsweek* mimicked *Time*'s Rocky Mountain boom story, but switched its geography to the Pacific Northwest, hyping Portland, Seattle, Boise, Bend, and even Spokane as "Pacific Northwest Paradises."[32]

Break from the Past

The New West expansion has some of the qualities of earlier natural resource booms: it attracts immigrants from all over the United States and the world, it creates tensions between the newcomers and longtime residents, and it threatens the integrity of regional ecosystems. Seemingly overnight, residents of some quiet western towns have found themselves living in ritzy resorts, knocked down a rung or two on the economic ladder by newcomers who build multimillion-dollar homes.[33] The expansion has environmentalists debating which is better: an overgrazed ranch or a subdivided ranch?[34]

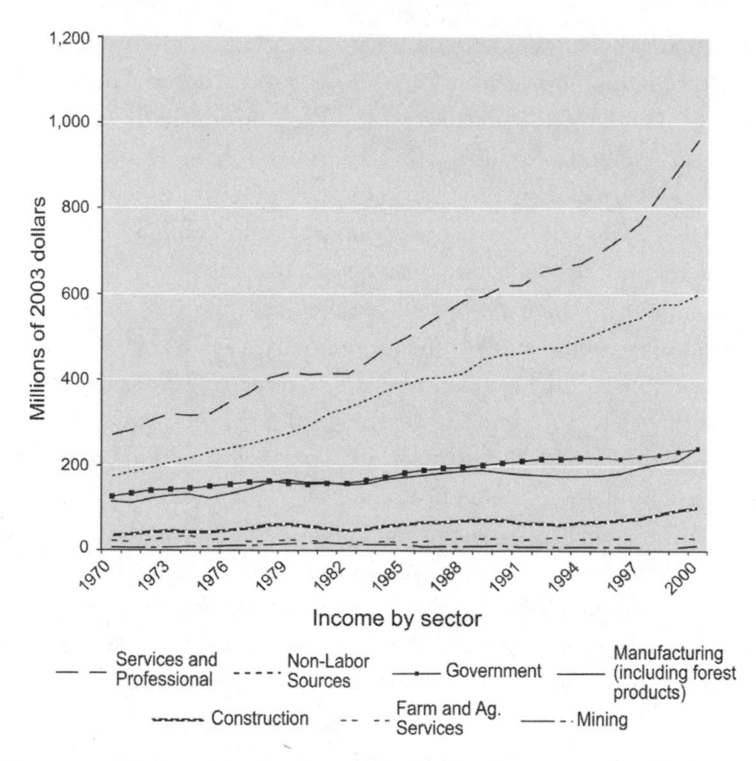

Figure 1.3 Personal income by sector in the eleven western states, 1970–2000. Although attention focused on the rapid growth of the New West economy in the 1990s, service and professional income has outpaced income from all other sources since the 1970s. Even more striking is the second-place showing for "non-labor" sources such as investments, rents, and retirement income. Manufacturing, government employment, and construction also grew, while agriculture, and mining lagged, constituting a decreasing proportion of the income pie. *(Data from the Bureau of Economic Analysis, analyzed by the Sonoran Institute's Economic Profile System, http://www.sonoran.org/programs/eps/si_se_epsindex .html, and provided by Ray Rasker of Headwaters Economics, see http://www. headwaterseconomics.org/index.php.)*

But current growth is also different from the region's historic economic patterns. Both cities and deeply rural places are experiencing dramatic new investment and growth. The most obvious characteristic that differentiates this boom from those of the past is the diverse set of high-tech, telecom, and service industries that are its foundation. Job growth was flat in all the traditional western industries in the 1990s (indeed, most extractive industries lost jobs), yet the region gained jobs, on a per capita basis, faster than the nation as a whole, and its population growth was the fastest in the nation as well.[35] Most of those new jobs, and the income they generate, show up in the category economists refer to as "services" (fig. 1.3), which brings to mind fast-food joints and grocery clerks. Such jobs certainly multiplied, along with a host of other service jobs associated with recreation, tourism, and retirement—activities that bring more people, with many needs, to the West.

But are service jobs the essence of the region's emerging economy? Economist Ray Rasker, in an article evocatively titled "Your Next Job Will Be in Services. Should You Be Worried?" answers, emphatically, *no*.[36] He points out that the "services" category is very broad, ranging from burger flippers to architects, from ski instructors to software developers, and that the sector includes the only high-paying and high-quality jobs actually growing in numbers. Rasker and others, who have come to constitute the "New West School," argue that the postextractive and postindustrial economy is driven by individuals and firms that can operate wherever they wish. Companies need not locate near a hydroelectric dam, coal seam, or seaport (although a commercial airport is important). They don't even need to locate near one another anymore, as telecommunications, transportation improvements, and changes in the nature of their products have eased the "friction of distance" and reduced the need for spatial proximity to suppliers. Thus, technoburbs sprout on the edges of Denver and Phoenix, and high-tech plants rise from wheatfields halfway between Denver and Fort Collins or from rangelands south of Reno.

Leaders in everything from biotechnology to robotics repeatedly cite the two most important characteristics of communities in which they will locate: a highly educated workforce and high quality of life. Likewise, economist Power argues that businesses and people locate in the West not because of some traditional economic attraction, but because of environmental quality:

> People care where they live and, given the choice, gravitate toward more desirable residential areas. They want high quality of life, including access to the West's natural landscapes and ecosystems. Economic activity tends to follow them. Environmental quality has become a central element of local economic bases and a central determinant of local economic vitality.[37]

This geographically footloose quality is what enables people and the economy to reshape the western landscape, and that reshaping is the topic of this book.

Enduring Growth?

The West's modern economic growth cannot be analyzed like previous expansions, and it should not even be called a "boom" (although the word is difficult to avoid). No single commodity, or even suite of related commodities (such as uranium and other strategic metals), can be identified as the driving force of the current expansion. This growth is not focused on a particular resource nor anchored in the subterranean geography of oil deposits or mineral veins. It is not occasioned by a singular discovery or by a political-economic event like the 1970s oil crisis. Instead, growth in the West reflects a maturing and diverse economy that combines elements of technology, investment, entrepreneurialism, and inherent regional qualities and hints at a sustained regional florescence.

Thomas Power and Richard Barrett argue that the West has entered an era of "post-cowboy economics":

> The economic vitality of the Mountain West does not appear to be tied to the same kinds of passing economic enthusiasms that gave rise to the gold rushes, land bubbles, and energy booms of the past. Instead, it seems as firmly grounded in permanent settlement as was development during the last half of the twentieth century in Southern California, Florida and Texas, and the rest of the sunbelt.[38]

The "wide variety of non-goods-producing activities" that now dominate the region's economy "can take place almost anywhere, particularly as the costs of communications and transportation fall."[39] The West, especially the interior, appears well suited to capture the new economy; its attractiveness to people and businesses is likely to enhance its growth for the foreseeable future.

Power and Barrett conclude that "amenity-driven relocation of economic activity appears to be an enduring force shaping the region's future."[40] Recent surveys of newcomers in California and Montana support the notion that amenities drive migration.[41] I give more attention to this process, and its implications for the pattern of future development, in chapter 2, examining the driving forces behind western growth.

Finally, it is worth pointing out again that the West still holds most of the nation's domestic energy reserves (both fossil fuels and renewable resources: the region is beset with plentiful sunshine and wind). The great irony is that higher energy prices, seen by some analysts as a potential brake on suburban expansion, actually drive more growth in the West, in both rural and urban settings.

As the West's economy loses its insularity and becomes more integrated with national and global trends, it will oscillate with broader economic swings. The great question for the New West School is whether such swings will be as sharp as past ones, or whether the greater economic diversity of the new regime will lessen their intensity. Simple theory dictates, and several regional analysts believe, that as the West becomes less vulnerable to price swings of one or two commodities, it will achieve greater economic stability.[42] Indeed, with environment and quality of life fueling regional growth, the one force that might defeat the region's success in the postindustrial economy is, simply, a loss of that regional charisma. In this vein, *Time* magazine's cover montage of the Maroon Bells and Denver's skyline reflects the conflict between development and environment, between livable communities and landscape-degrading sprawl, a tension that has intensified considerably during the last few decades. Although mostly upbeat, *Time*'s take on regional development did hint at the costs of growth. Those costs would be loudly articulated by the anti-growth movement that had emerged across the West by the end of the 1990s.

The slowing economy in the early 2000s illuminated the abiding challenge that growth management efforts face. Only a few months into the 2001 economic slowdown (reflecting the national recession that lasted, technically, only from March through November of 2001[43]), political leaders began to block even the limited, hard-won progress toward growth management elicited by the 1990s boom, arguing that the West now needed growth promotion rather than growth management. On the first day of a special (September 2001) session of the Colorado legislature, the *Denver Post* reported that "the debate over what to do about Colorado's booming growth, which was expected to dominate the special session, suddenly has gone bust."[44] Politicians wanted to shelve plans that would curb urban sprawl or in any way send a message that Colorado was not pro-growth. Some legislators argued that the earlier enthusiasm for growth limits had actually caused the slowdown! The *Denver Post* quoted one state representative as saying that "any time government starts talking about placing heavy regulations on industry, any industry, the economy is going to be impacted."[45] Growth management is thus caught in a conundrum: it is slow to emerge during growth spurts, and it is quickly discarded in slower economic times. But this slowdown did not last; indeed, the region's history is marked more by growth than by contraction, and the houses, highways, office parks, and resorts added to the western landscape will remain, accumulating, faster or slower according to the economy, in a permanent lessening of open space and wildlife habitat.

This latest real estate boom is also transforming the social landscape in unhealthy ways. Upward-spiraling property values in resort areas create stark class divisions.[46] Inappropriate growth, big-box retail development, and low-density exurban development on the outskirts of small towns disrupt the easy functionality and social fabric of those towns. In addition to invading large swaths of agricultural and natural lands, these developments also cost more in public services than they contribute to local tax coffers.[47] Even the West's suburbs—little appreciated by urbanists and criticized as inefficient and lacking soul—offer a scale and functionality that is now threatened by hyper-development, beltways, gated communities, and mega-malls.[48] Each beltway and suburban-edge development encourages and subsidizes investment farther out, extracting wealth from older suburbs and urban cores and cutting existing suburbs off from the open spaces that so many westerners value.

Future Landscapes

The "long boom" interpretation of western history suggests that more development is on tap. How much more? Although we can hardly expect the growth rates of the 1990s and early 2000s to persist forever, I believe that the region's population and economic growth are poised to outpace the nation's for decades to come. I argue this point throughout the book, but let me anticipate here some old-fashioned notions of geography that seem to have westerners constantly expecting regional decline. As a geographer, I was trained to believe that physical landscapes and natural resources place some limits on development. The West's dry climate would seem to place an absolute limit on population growth, and its big spaces, its geographic barriers such as mountain ranges and canyons, and its widely separated cities should, in theory, retard development. The "friction of distance" should make commerce of any type slow and costly.

But instead of limiting growth, the West's large swaths of public lands and its dramatic mountain, canyon, and desert terrain now *attract* and encourage development, from the desert golf communities in Arizona and California to ranchettes in Montana and Wyoming. Newcomers, including CEOs who bring entire companies with them, cite the West's landscape and outdoor lifestyle as reasons for locating there.[49] Even the region's aridity, long assumed to be the preeminent limit on western development, seems to have lost its power to retard growth. Enormous dams, tunnels, and canals that collect and move water across mountain ranges and through deserts, built originally for agriculture, now enable the spread of residential and commercial land uses throughout the region. If anything, there is too much water in the West, so much irrigation water (some 80–90 percent of water use in the region is still in agriculture) looking for industrial and municipal buyers in a poor agricultural market that we have annihilated the West's aridity, or at least the role it might have played in limiting development. Rather than water attracting growth—or lack of water limiting it—growth itself attracts water.[50] Even the worst-case future drought can be managed by the region's adaptable, interconnected water systems.[51]

Finally, our changing attitude toward wilderness (once dreaded, now loved; once only a place to visit or exploit, now a place in which to live or, at least, to live near) encourages western development.[52] Instead of a foreboding, coarse landscape that resists settlement, we have made the West into a frictionless geography that welcomes settlement and development, even, perhaps especially, in its more remote, natural reaches. Timothy Duane, dissecting the growth he believes is hurting the Sierra Nevada, called the widespread settlement that accelerated in recent decades the "eco-transformation" of the social and physical landscape.[53]

2 Development Geographies of the New West

THE LONG TRAJECTORY of growth in the West, quickening over the last two decades, has plastered multiple development patterns onto the natural topography. How do we make sense of these complex landscapes and landscape changes? Many of our conceptions—our mental maps—of the region are outdated, like my daughter's grade-school atlas that bedecks the West with oil derricks, copper mines, and a couple of ski areas instead of the high-tech office parks, sprawling suburbs, and hobby ranches that now take pride of place in the contemporary western landscape. We need an updated geographic template to measure the West's development landscapes, one that speaks not only to planners and ecologists, but also to the growing population of westerners who worry that their communities and landscapes are changing fast—too fast—in ways they had not anticipated and, on the whole, don't much like. In this chapter, rather than dividing space into discrete physical and cultural units as in the geographic tradition, I offer a typology of "development geographies" layered on, and interwoven with, the western terrain.

Four Geographies

Four distinctive land use patterns now dominate western development. I describe each briefly here and then take them up in detail in part 2.

Metro-Zones

The West hosts the fastest-growing cities in the United States. In nighttime satellite photographs, some, such as Phoenix and Las Vegas, stand out as circles of light against a black background of open spaces (fig. 2.1). Others sprout multiple tendrils that connect peripheral blobs of light in lopsided constellations, such as those across Southern and Northern California and northwestern Washington. Broad stripes of light mark urban corridors, stretched out along mountain fronts (such as Utah's booming Wasatch Front and the Colorado Front Range) and along rivers (such as the string of cities growing along the Snake River

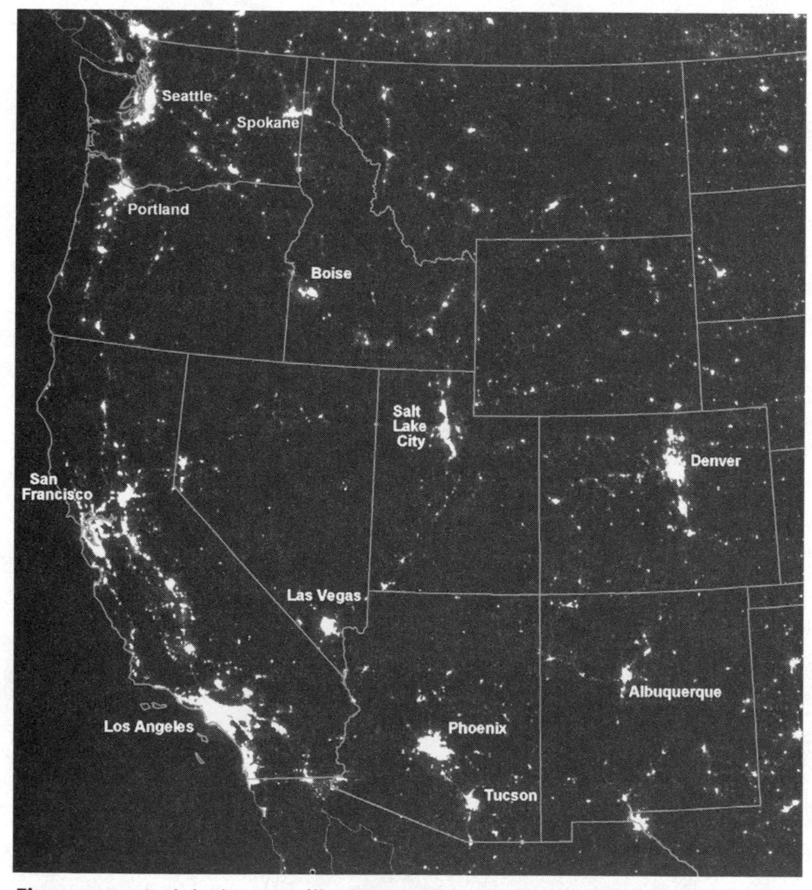

Figure 2.1 A nighttime satellite image of the West reveals a growing network of development against a dark background of open space. *(National Oceanic and Atmospheric Administration photograph.)*

Figure 2.2 Though Las Vegas is one of the more concentrated cities in the West, even its suburbs inevitably sprawl out onto the region's signature open spaces. *(Landslides Aerial Photography photograph.)*

in southern Idaho and astride the Rio Grande in North Central New Mexico). But "city" no longer serves to describe the West's urban geography. Rather, the West is marked by sprawling *metro-zones*, made up of multiple cities, suburban development in the unincorporated interstices among the cities, and open spaces now subsumed into the metropolitan landscape. Even smaller cities—such as Bozeman, Flagstaff, and Bend—have sprouted new residential and commercial developments out past their edges, spurring the U.S. Census Bureau to create a new category—"micropolitan" areas—that recognizes these growing cities as metro areas in their own right.[1]

Western analysts like to point out that the region is markedly "urban" despite its agrarian varnish. The vast majority of westerners, including most of those who arrived in the recent surges of population growth, live in suburban settings (fig. 2.2) (some 82 percent in 2000, compared with a national average of 78 percent). The region's urban-focused settlement pattern is not new: even during early settlement spurts, including the gold rushes, cities dominated the West's economic and social landscape. Although serious historians have documented this part of the West's history, the region's frontier creation myth largely ignores it.[2]

What does the West's urban character mean geographically? Most westerners are not technically urbanites, but suburbanites, and the suburbs, not the central cities, have seen most of the region's growth since World War II and have gained most of the new jobs and businesses. Indeed, the cities are dominated by the suburbs geographically, visually, and more and more, politically and economically. Suburbs will absorb the majority of the 28 million additional people expected to live in the eleven western states by 2030. Although you may hear that Phoenix and Las Vegas are booming, their newer suburban cities, with less famous names, are actually growing the most rapidly. During 1990–1998, the Phoenix suburb of Chandler grew by 78 percent (adding 70,000 residents); Henderson, outside Las Vegas, more than doubled in population; and Mesquite, Nevada, grew from 1,821 residents into a fair-sized suburban town of 10,125—an increase of 441 percent! These suburbs, and many others in the West, such as West Valley City (southwest of downtown Salt Lake City) and Peoria (on the edge on Phoenix), are "edge cities": suburban areas that act as satellite residential and business districts to the older, core cities. But some urban analysts are beginning to doubt the necessity of a prime city as the economic and political power of suburban cities grows and they build their own highways, airports, and water supply systems. The future of western cities may be suburbs without the "urbs."[3]

The most intense land transformation in the West occurs in its cities, in both cores and suburbs. Pavement or buildings cover most of the land in the cores, inner suburbs, retail strips, and edge cities, with hundreds of dwelling units or thousands of square feet of office space per acre. But in the new suburbs, residential subdivisions, low-rise office campuses, and retail land uses are much less dense, taking up to ten times the land area for the same density of dwelling units and office space as in the core. In addition, because they encompass more parks and open space, many suburbs exhibit more seminatural landscaping than hard surfaces. But these heavily manipulated "natural" land covers can in only the most constrained way be counted as part of the West's natural landscape: they provide little wildlife habitat, except for the most human-adapted species, and they don't react to energy and water inputs in the same way that truly natural surfaces do.

The development footprint of western cities grows faster than their population; that is, they sprawl. There is little geographic reason to expect this pattern to be reversed. Most western cities appear to have practicable plans for land and water resources that will support decades of continued growth, at decreasing overall population densities, even as their citizens complain about traffic, lost views, and crowded schools. Local barriers to urban growth—a mountain range here, an arc of public land there—are just that, local barriers, soon to be flanked or leapfrogged. Residents of interior places such as Denver and Phoenix complain that they do not want their cities to become like Los Angeles. But what sociologist Harvey Molotch called the "growth machine"[4] and historian Hal Rothman more gently calls the "growth coalition"[5]—the convergence of residential preferences, household finances, government boosterism, and corporate strategy that keeps cities and suburbs growing—has those cities headed precisely down Los Angeles's path to megalopolis.

The high-tech edge-city office parks, beltways, and new mega-airports of the Interior West are emblematic of the same forces that built Los Angeles, Atlanta, Chicago, and Dallas–Fort Worth, forces that operate as effectively in the West as elsewhere. It may be difficult to imagine 10 million people living in Colorado's Front Range metropolitan corridor, but it was equally difficult to imagine today's Los Angeles of 16 million inhabitants only fifty years ago, when the county held 4.1 million, or in 1900, when only 170,000 people called it home. Denver leaders did not fight to build the nation's biggest airport to limit growth, and with irrigation water flooding the agricultural lands around them, Front Range cities do not consider water supply as a limit to growth. Indeed, western cities face few fixed limits on growth.

The Exurbs

The cities provide an economic and geographic tether to an emerging *exurbia*, the next step out from suburbia. Part rural and part suburban, the exurbs are well removed from cities, with dispersed, low-density, residential land uses and pockets of commercial development (fig. 2.3). As companies move to the suburban edge and governments improve roads and infrastructure,[6] people can live well away from the city and maintain a tolerable, mostly rural, commute to work. Self-employed and retired exurbanites—not tethered to the city at all—also fuel demand for exurban residences and associated development. Real estate companies have figured out the appeal of the exurbs, and their ads promise exactly what exurbanites are seeking: great views of mountains (and maybe distant city lights), elk outside your window, neighboring public lands on which you can roam, and all within, say, an hour's drive to city, airport, and ski slope. The appeal is great, and the exurbs are growing fast. John Cromartie, a geographer with the U.S. Department of Agriculture, has shown that nonmetropolitan counties adjacent to metropolitan areas are the fastest-growing places in the United States (think Park County, Colorado, in the mountains southwest of Denver, or the Sonoita Valley, well outside of Tucson: both long, but apparently not unacceptable, rural commutes from city

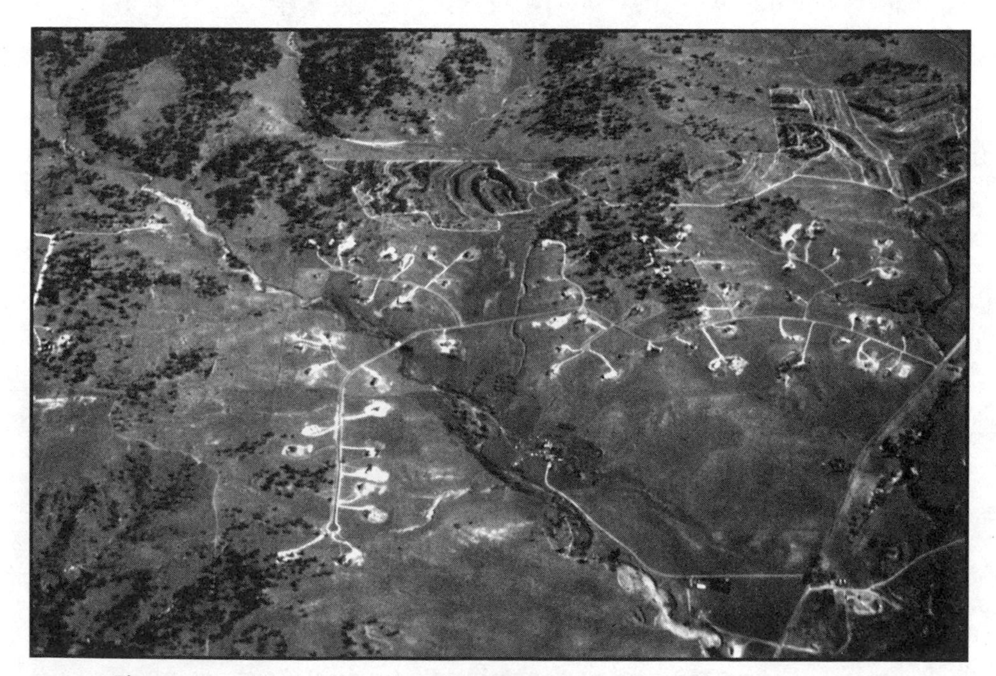

Figure 2.3 The exurbs: a large-lot development carved from a ranch in the foothills southwest of Denver, Colorado. *(William Travis photograph.)*

jobs and services).[7] The western exurbs also gained jobs faster (44.5 percent) than metropolitan (26.6 percent) or more rural counties (32.5 percent) during the 1980s and 1990s.[8] The exurbs are really part of the metropolitan geography of the West—their populations rely on the nearby cities, if not for daily work, then for urban services: hub airports, major hospitals, entertainment, venture capital, banking, and even city lights—in the distance.

The exurbs now cover as large an area of the western landscape as do the cities, and are spreading faster, albeit with a less solid footprint (with residential densities of one dwelling unit per 5–40 acres). The exurban landscape is the scene of the West's greatest tensions between people and nature. Here wildfires, wildlife, and people uneasily, sometimes dangerously, coexist. The effort to suppress wildfires in the exurbs incurs large ecological and monetary costs (and risk to firefighters), and those elk touted by the real estate ads damage landscaping, frighten residents, and attract predators. This new residential frontier, the exurban tension zone, demands close examination.

The Resort Zones

Another, especially conspicuous development landscape type, the *resort zone,* perforates the West in places such as Aspen, Jackson, Palm Springs, and Sedona. This kind of development, driven by recreation and tourism, brings extravagant new commercial and residential investment to mountain and desert environs, creating landscapes that journalist Raye Ringholz called "paradise paved" and author Ed Abbey disparaged as industrial tourism.[9] Ringholz wrote of small ranching and mining towns discovered and transformed into resorts, their residents overwhelmed by an "influx of new and often wealthy residents" that drives up property values and "threatens to eliminate the qualities that made these places attractive in the first place." She wrote: "What was to become the cliché of 'Aspenization' reached into emerging villages like Moab, Utah, and Ketchum, Idaho, where the pinch was just starting to be felt."[10] My own travels to these towns and others found many overwhelmed by the spreading geography of resorts (fig. 2.4).

Historian Hal Rothman analyzed resorts, from Sun Valley to Squaw Valley, as a new form of corporate-controlled colonialism, not unlike mining and energy development.[11] The development (base villages,

Figure 2.4 The mountain resort's footprint includes base development and ski slopes reaching to the summits. *(William Travis photograph.)*

massive homes, condos, swank retail, high-end services, and golf courses) and its secondary effects (such as shortages of affordable housing and distant resort-worker bedroom communities lacking social services) bring a set of poorly dealt with land use problems to the West's most natural and wild landscapes. The towns and recreational villages created in the resort zones are like mini-downtowns, with high-rise hotels and concentrated retail and conference centers, often pressed right up against wilderness. The second-home subdivisions expand these effects outward into residential landscapes that stick out visually because of the large sizes of the homes and their "view"—often ridgetop—locations. Farther out are the small towns and new residential enclaves filling with resort workers who can't afford housing in the resort town proper. As the resort zone expands, these worker residences themselves become Aspenized, and the workers are pushed farther away from the resort area.[12]

The Gentrified Range

Beyond the exurbs and resorts lies the *gentrified range*, a fundamentally rural landscape, but one now dotted with hobby ranches and other New West homesteads. The gentrified range is something of a terra incognita, which makes it hard to analyze as a discrete landscape type. Some of it is akin to the exurbs of small to medium-sized western towns (however, I distinguish it from the exurbs because it is not within commuting distance of the city), and nearby resort zones certainly add to the interest in and growth of amenity ranches and ranchettes (fig. 2.5).

In traditional rural studies, population increases in rural areas meant only one thing: an expanding farm, ranch, timber, or mining economy. Now those economies are flat or dying in the rural West, yet its population is still increasing; indeed, the population of rural counties in the West grew an average of 20.7 percent, a bit faster than the rate of 19.6 percent in metropolitan counties, in the 1990s.[13] These new rural settlers bring their jobs and incomes with them, and they demand services not typical to rural economies (creating what demographers Kristopher Rengert and Robert Lang called "Cappuccino Counties"[14]).

The landscape effects of this rural gentrification are subtle. Typical

Figure 2.5 The gentrified range. Working ranches are increasingly for sale in charismatic landscapes across the West, marketed as private retreats. *(Julia Haggerty photograph.)*

low-density development places houses, roads, and other facilities on former farms and ranches at densities akin to those of exurbia. But some of the change is more about land management than development, as people with little connection to traditional agriculture buy and manage some of the West's largest ranches. One recent study suggested that roughly half of the West's ranches are now "hobby" operations, owned mainly for their landscape amenities and investment value rather than for livestock production.[15] Besides isolated homes, new horse arenas, and maybe even an airstrip and hangar, the transition in ranchland ownership brings far-reaching changes in land use as the new owners implement their own ideas on grazing, wildlife, water use, and access. This rural gentrification—what some geographers studying rural change in Britain call "greentrification" as newcomers seek to preserve natural landscapes[16]—also changes the politics, economics, and culture of western rural areas in quite profound, but poorly studied, ways.[17] Although traditional ranchers will disagree, it may be that the new interest in ranchland amenities brings more support for wildlife and habitat preservation to rural places.

Developing Landscapes

These four dominant development geographies capture the most important contemporary patterns, but this typology is not the only way to divide up western development, and it does leave some gaps. Certainly the West's tribal lands deserve more attention, and although many good studies of Native American history, culture, and policy exist, I am unaware of any comprehensive studies of contemporary reservation land use and development patterns (a good project for the right Indian Country expert). My four geographies also miss the small towns that don't logically fall into the rural or resort categories. Some are booming, growing into what the federal Office of Management and Budget and the Census Bureau have begun to call "micropolitan" areas. A few small towns are still tightly linked to the natural resource economy, such as Rock Springs and Rawlins, Wyoming, both of which lost population in the 1990s but started gaining again in the early 2000s as another wave of energy development swept the region.

In addition, much of the West is public land, subject to different types of development than I analyze in this study. The federal lands—the focus of decades of argument about natural resource and land use[18]—offer an increasingly stark contrast to private lands. Although clear-cuts, roadless areas, and oil and gas drilling on federal lands will certainly continue to be matters for heated debate, the private land development that is creeping up to the boundaries of public land and creating a sharp divide between the most natural and the most developed landscapes in the region will increasingly draw westerners' attention.[19] In addition, development on private lands affects the management of nearby federal lands: public land managers must work even harder to suppress fires and to aggressively manage the wildlife that often strays across public-private boundaries. Thus changes on private land further compromise the naturalness of federal lands and place their biodiversity at greater risk.

The development geographies I describe here are not uniquely western; most of these patterns replicate themselves across the American landscape. Urban sprawl in Atlanta is similar to what it is in Phoenix, and the western resorts have equivalents in places such as Saratoga Springs and Sanibel Island (and perhaps can learn some lessons from those older American resorts). The term "exurb" was coined to describe development outside the Mid-Atlantic megalopolis, so the western exurbs are not unique either. In the West, however—a region of relatively few urban areas surrounded by extensive wild lands—exurbia more dramatically alters long-standing connections between city and countryside and muddies the relationship, often already tense, between the urban and agrarian subcultures tied to those landscapes.

~ ~ ~

I can see examples of most of these development landscapes through my home-office windows as I write this. To the west, on the slopes of the Front Range, white streaks of ski runs stand out on a mountain backdrop that is mostly federal land. The lower foothills are dotted with houses, whose presence is signaled in the morning as the rising sun glints off their large picture windows. In the foreground, the suburbs of Boulder spread across a few miles of High Plains prairie. Boulder's suburbs are pretty much like those of any city, except in one important way: their outward edge is fixed in perpetuity by the most extensive municipal open space system in the nation. Looking south from the hill behind my house, I can see high-rises in Denver, surrounded by a ring of suburban sprawl and edge cities, separated from Boulder by the open space purchased to do just that: separate the sprawls. Finally, to the east, traditional and hobby farms and ranches, horse properties, commercialized road and highway intersections, and isolated big houses on large lots dot the exurban Plains, all against the backdrop of Denver's new mega-airport shining in the setting sun.

This arc of landscapes is visible from my house because I live adjacent to some of Boulder's open space, open land protected partly because, in the early 1970s, my leapfrog subdivision crept far enough up Gun Barrel Hill—adjacent to a valued natural area—to be visible from downtown Boulder. The city and country responded with one of the country's most aggressive comprehensive plans and one of its most effective open space programs, meant to check such spreading development. Indeed, I might be looking out on a landscape rare in the modern West: one on the cusp of regulated buildout, a place about to enter (through a variety of planning tools) the unexplored country of no additional significant greenfields development (that is, conversion of raw land to subdivisions or commercial uses). Yet Boulder stands out in western land use planning as an island of restraint in a sea of unbridled development; cases of such growth management and land use planning are the exception rather the rule in the West.

Making the New Geographies: Driving, Enabling, and Shaping Forces

Whether universal or regional, the outcomes of land development occur in real places and are fixed in space. People who care about the Rockies, the Colorado Plateau, the Sonoran Desert, and other distinctive western landscapes have good reason to critically examine the development they see out their windows and to consider whether they can alter its patterns to create more desirable and more sustainable geographies. But if westerners are to alter the trajectory of regional development, they must grasp the social factors that propel it. Both long-standing and more recent forces in the American social and economic system shape the new geographies of the West: the exurbs of Denver and Tucson, the gated communities of Vail and Santa Fe, the office parks south of Reno and north of Colorado Springs, and the ranchettes and hobby ranches commandeering western Montana's broad valleys and Arizona's southeastern grasslands. Federal laws and policies are often given pride of place in western studies, but to understand the New West's development landscapes, we must look to local actions as well, to towns and counties that pass (or fail to pass) land use ordinances, create master plans, build roads, buy open space and wildlife habitat, or try to fill their tax coffers by pursuing another big-box retailer or software company. And we must attend to the great engine of private capital accumulation and its deployment, especially via real estate speculation and investment. Finally, we must grapple with human nature: exactly how much of what we see in the developed landscape exists because of personal preferences? Answers to these questions are important in devising effective tools for altering the trajectory of regional development, for reducing its negatives and accentuating its positives.

To make sense of the forces behind the patterns on the ground, we need a second organizing scheme. The development landscapes explored in the following chapters arise from a coalescence of three types of forces—driving, enabling, and shaping forces—that mold capital, preferences, and regulations to create western development as we witness it. National and global economic forces, international immigration, and regional population growth all drive western development. A burgeoning love of western landscapes and long-standing American preferences for residential space (which drove the postwar suburban boom) further fuel the West's new sprawl.[20] Income growth and corporate mobility, employment flexibility, retirement trends, and communications, transportation, and construction improvements enable the new western homesteading, especially its exurban and rural arms. By the end of the twentieth century, these forces were enabling what Hal Rothman and others called the "Re-Opening of the American West," a postindustrial, postmodern second rush to the frontier.[21] Finally, terrain, government regulation, and the public lands all shape western settlement.

A group of economists, geographers, and demographers—whom I label the "New West School"—are grappling with the causes and consequences of rapid growth and development in the West.

Although the New West School throws together all the causes of western growth under "driving" forces, I more specifically define each underlying force to more clearly link it to different types of actions that can alter growth trajectories. But driving, enabling, and shaping forces can blur together, making it difficult to make clear distinctions among them. Some of the macro-level forces that New West School analysts cite—such as non-earnings income associated with investments and retirement as well as diversifying economic opportunities overall—act as enabling factors because they allow people to act on their preferences. However they are defined, each of the causes behind the new patterns of place emerging in the West—driving, enabling, or shaping—deserves our attention; if we are to deflect the current trajectory of regional development, we must alter one or more of them.

Driving Forces

I begin with driving forces because they are the "why" of development. Why is the West growing so fast? Why has it become the destination of choice for footloose industries, small businesses, high-tech firms, international and domestic migrants, and retirees? Why do its cities sprawl? What is driving the run on ranch real estate?

Most explanations of regional growth start with population growth, but the real driving forces come one step before population growth. They include the preferences that affect where people live, the economic and political forces driving immigration, and the demographic forces that govern in situ population growth. The New West School focuses on people's preferences as a force behind population growth. They are impressed with evidence that many—maybe most—Americans prefer small-town and rural life if they can get it. The West—even the West's cities—seems to offer this lifestyle. Geographers William Beyers and Peter Nelson argue that people are increasingly able to act on this preference because the West's economy, even in rural areas and small towns, has diversified. Increasing value attached to outdoor amenities also drives both urban and nonurban growth in the West.[22] The in-migration process reinforces itself, they contend, because amenity-seeking new arrivals create further economic opportunities for others in preferred locations.[23]

Population Growth Regional development theory is full of chicken-or-egg questions: Do home-buyer preferences, housing supply, or zoning regulations create the sprawling repetitiveness of modern suburbs? Do growth limits or the innate attractiveness of places that eventually feel compelled to enact growth limits drive up property values? One solution to a chicken-or-egg conundrum is simply to pick an obvious starting place and work from there, and for most analysts explaining the region's growth, that starting place, taken here for argument's sake as the primary "chicken," is population growth, especially through immigration. Both components of immigration to the West—domestic and international migration—are strongly positive (meaning that more people arrive than leave). During most of the 1990s, for example, the Rocky Mountain states attracted more immigrants from all other census regions than they sent back. Even California showed a net out-migration to the Interior West for much of the 1990s.[24] But California still grew because of net international immigration (and due to high fertility, as we will see shortly).

International immigration is a large, and controversial, driving force in western development. Although data on immigration are poor, we do know that most of the roughly 1 million documented immigrants to the United States each year settle in the West: California alone receives roughly a quarter of a million documented immigrants yearly. Another 700,000 or so (perhaps significantly more) undocumented immigrants also enter the country each year, mostly in the West, and many stay permanently. Mexican citizens (and increasingly citizens from other countries in Latin America), driven north into the United States by economic and political factors, dominate this migration, which now ranks with growth as a popular topic of op-ed columns, letters to the editor, and political debates in the West. Critics attribute a variety of social ills to international immigration, but for the purposes of this look at western development, we have only to note that it helps to fuel population growth in the West and provides a pool of labor, both skilled and unskilled, that bolsters the region's economic growth. Indeed, many proposed legislative fixes for illegal immigration include some form of amnesty and, eventually, citizenship. Such measures recognize the value of immigrants to the economy and, of course, would probably spur further population growth.

It is also worth noting that immigrants tend to be relatively young and exhibit relatively high fertility, thus ensuring future natural population increase in the region.[25] Immigrants also build wealth, becoming consumers and homeowners. In this way, the American West is establishing a demo-economic profile that other regions and countries—especially western Europe and Japan—are starting to envy as their internal fertility rates decline amid an aging population. A postindustrial economy fueled by immigration, both domestic and international, may well be the best model for continued growth in the early and mid-twenty-first century.

The West also exhibits relatively high rates of "natural" population growth, meaning births over deaths. Its fertility rates are well above the national norm: the eleven western states average 71 births per 1,000 women of childbearing age, with Utah topping the chart at 92 births per 1,000 (compared with the national average of 66).[26] This keeps the West relatively young, as evidenced by the broad-based current and projected population pyramids for most western states (fig. 2.6). But the region also attracts older domestic migrants (thus Arizona's bulge of both older and younger populations in fig. 2.6). The coming American retirement boom will affect more than the traditional retirement hot spots in the desert Southwest. When author David Savageau updated *Retirement Places Rated* in the mid-1990s, he found his surveys pointing to nontraditional retirement magnets such as Sandpoint, Idaho, Kalispell, Montana, and Fort Collins, Colorado (which showed up as the highest-rated retirement spot in the country).[27] He reported that baby boomers were retiring, or preparing to retire, away from the traditional California or Florida places where their parents found refuge. Because we are only at the start of this surge in retirement, we don't know exactly how it will play out geographically. One recent study predicted strong retirement growth in small- to medium-sized cities in the South and throughout the West, and certainly the current influx of retirees into towns such as St. George, Utah, and Fort Collins, Colorado, supports this prediction.[28] But retirement also expands the suburbs of Phoenix and Las Vegas.[29] With the United States on the verge of its biggest retirement surge in history (the U.S. population over 65 will grow from 31 million in 1990 to 53 million in 2020), retirees will continue to add their num-

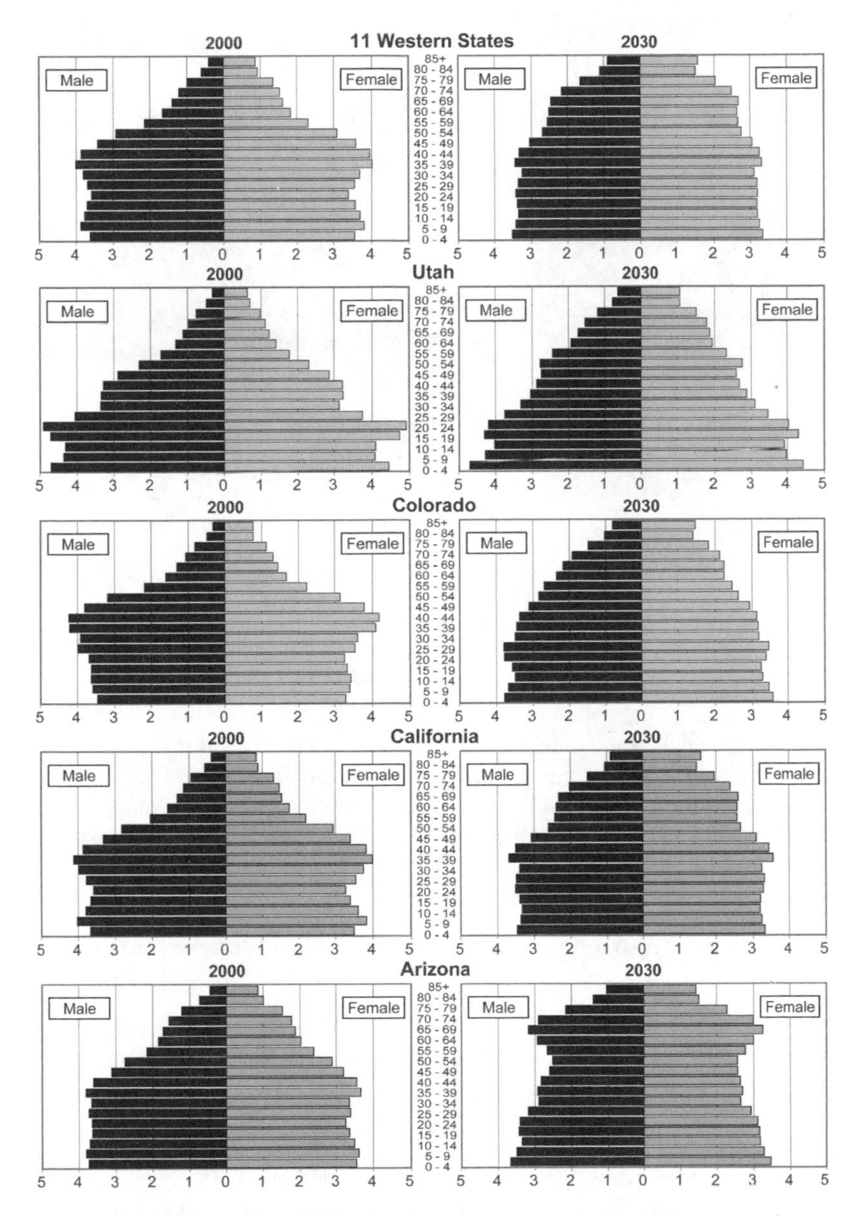

Figure 2.6 Population distribution by age group at the 2000 census and projected for 2030 for the eleven western states combined and for Utah, Colorado, California, and Arizona. A pyramidal shape indicates a growing population. Pyramids with large bases in 2030, such as those for Utah and Arizona, predict continued fast growth and high fertility well past the middle of this century. Overall, the West in 2030 should be even younger and faster growing than it was in 2000. *(U.S. Census Bureau.)*

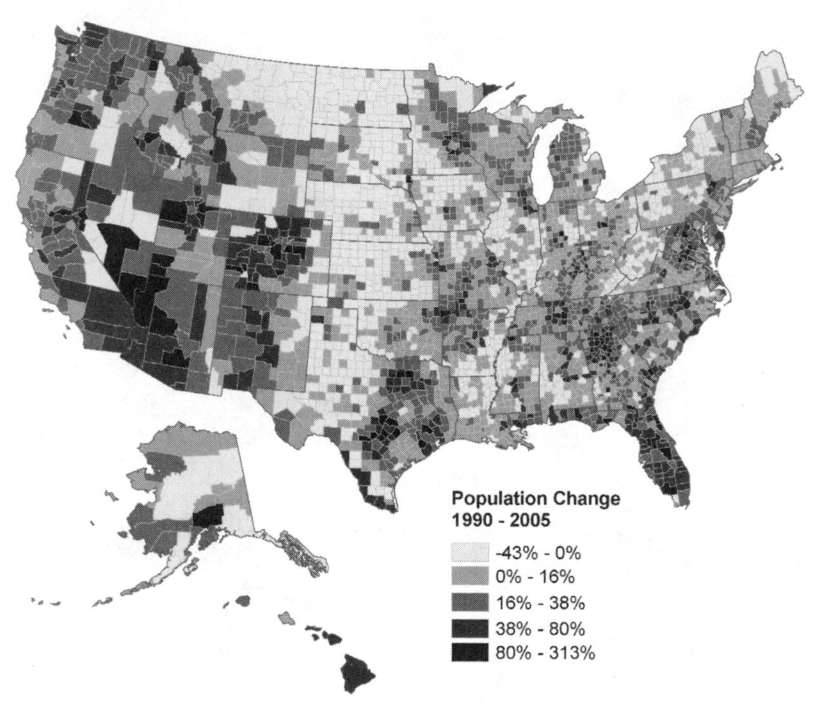

Figure 2.7 U.S. population growth rates, 1990–2005, by county. Pockets of growth are visible throughout the country, but the West stands out with a majority of fast-growing counties. The slowly growing and even declining swath of the Great Plains sharply delineates the eastern border of this growth zone. *(U.S. Census Bureau.)*

bers and wealth to a region already booming with immigrants spanning the economic spectrum.

Population growth has occurred almost everywhere in the Interior West. Away from the Great Plains (a swath of negative growth on any map of recent U.S. population change, and a region that, in essence, marks the eastern border of the "New" West; fig. 2.7), only thirteen counties showed flat or negative growth. These were deeply rural counties, but overall, western rural counties were among the nation's fastest-growing counties. Most cases of population loss in the Interior West result from a rather traditional cause: the decline or failure of a particular commodity or industry, such as the closed silver mines in North Idaho; loss of the lumber mill in Jackson County, Colorado; and the weakened mineral and ranching economy of Sweetwater and Carbon counties in southern Wyoming, where towns such as Green River and Rawlins saw almost 10 percent population loss in the 1990s.[30]

Thus the West presents a rare regional demographic profile: strong growth of all age cohorts and strong net positive immigration, both domestic and international. Indeed, many westerners sensed that their region was growing fast in the 1990s, and the 2000 census substantiated that feeling. The eleven western states grew by 10.2 million people in the 1990s, or by 20 percent (the national rate was 13.2 percent). This growth continues a historical trend that has put the West ahead of national growth rates for four consecutive decades. The Interior West topped the national charts of population growth, with Nevada, Arizona, Utah, Colorado, and Idaho making up the five fastest-growing states in the United States. These five states grew from 10.8 million to 14.9 million residents (4.1 million additional people, or a 37 percent increase: almost three times the national rate) during 1990–2000. The region's strength in all the components of population change is why most demographers expect it to lead the nation in growth for decades, and the Census Bureau's release in 2005 of population projections out to the year 2030 offered no real surprises for westerners. Current trends will continue, and the West will grow faster than the country as a whole, over the next three decades. Arizona and Nevada will more than double in population, and several Rocky Mountain states, such as Colorado, Utah, and Idaho, will add another one-third to one-half to their 2000 population (table 2.1). Even California, building on a huge base (33.9 million), will grow by more than one-third. In all, some 28 million more people will live in the West by 2030 than in 2000.[31]

Economic Growth By starting with population in our analysis of the driving forces of regional development, we risk falling into the truism "people cause growth." We also encounter another chicken-and-egg conundrum, one raised by the New West School: which comes first, population growth or job growth, which then attracts people? The traditional economic model put jobs first: jobs created by industries located in a place because of some geographic attraction, such as an ore deposit, hydroelectric dam, or transportation node. In this "base economy" model, people then move to the jobs, earn incomes exporting raw materials or manufactured products, and create an economic ecology of additional jobs and a tax base that round out the community. Most economic analysts and development directors accept this model and put most of their efforts into luring employers. The director of Colorado's Office of State Planning and Budget, alluding to (and dismissing) the notion that Colorado is attractive to immigrants even when job growth is slow, stated what almost all development officers fundamentally believe: "The only reason people will be coming is if there are jobs."[32]

Table 2.1

Recent and projected population growth in the eleven western states

State	1990	2000	% Growth 1900–2000	Projected Population 2030	% Growth 2000–2030	Total Growth 2000–2030
Nevada	1,201,833	1,998,257	66	4,282,102	114	2,283,845
Arizona	3,665,228	5,130,632	40	10,712,397	109	5,581,765
Utah	1,722,850	2,233,169	30	3,485,367	56	1,252,198
Idaho	1,006,749	1,293,953	29	1,969,624	52	675,671
Washington	4,866,692	5,894,121	21	8,624,801	46	2,730,680
Oregon	2,842,321	3,421,399	20	4,833,918	41	1,412,519
California	29,760,021	33,871,648	14	46,444,861	37	12,573,213
Colorado	3,294,394	4,301,261	31	5,792,357	35	1,491,096
Montana	799,065	902,195	13	1,044,898	16	142,703
New Mexico	1,515,069	1,819,046	20	2,099,708	15	280,662
Wyoming	453,588	493,782	9	522,979	6	29,197
TOTAL	51,127,810	61,359,463	20	89,813,012	46	28,453,549

Source: U.S. Census Bureau.

A year later, Colorado's state demographer, noting a falloff in jobs, told the *Rocky Mountain News*: "What's sort of remarkable is the fact that we have lost a heck of a lot of jobs, and we still have people coming in."[33] Still later, the *Denver Post* quoted the state's new demographer:

> That we could lose 100,000 jobs from 2001 to 2003 and still see people moving here totally surprised us. . . . It's completely contrary to the idea that people will follow jobs.[34]

Analysts in the New West School doubt the simple jobs-then-population equation.[35] They argue that in an economy now dominated by services, quality of life attracts people as much as jobs do, and that people bring, attract, or create jobs in situ. I see their argument not so much as an outright challenge to a logical supposition (that jobs attract people), but as a quarrel with the jobs-at-any-cost mentality so common across the region.[36] People care where they live, and in an increasingly footloose economy, many no longer slavishly follow jobs.[37] The New West theorists believe that landscape amenities now have a large effect on where people *and* jobs locate, and that development proponents, when they finally internalize this paradigm, will work to protect environmental as well as economic amenities.

The New West School has pursued and strengthened this argument for a decade now, and a growing literature backs it up in the case of small towns, resorts, and rural areas in the West.[38] The argument has been most thoroughly examined in the area around Yellowstone National Park, where, for example, sociologist Patrick Jobes has studied immigrants for over twenty years. Jobes fills an entire book detailing the causes and effects of migration to western Montana.[39] He tries to disentangle economic and social aspects of the area's rapid small-town and rural population growth, finding most immigrants driven by a vision of living in "a beautiful, safe, and somewhat remote place," what he calls "moving nearer to heaven." That vision is decidedly noneconomic, maybe even irrational, and many of Jobes's subjects were eventually disappointed; half were gone in five years or less. But others stuck it out, adjusting to lower incomes to live in an attractive ecological and social setting. Jobes does not reject economic theories of migration, but he does question their monothetic explanatory power and finds, at least in western Montana, that noneconomic goals dominate people's choices of where to live. A few other studies have found a willingness to trade off salary or cost of living for quality of life in high-amenity areas.[40] Attractive, recreation-rich areas such as Yellowstone tug at those who have visited them, and it makes sense that they would bear out the New West School's argument. One survey revealed that many business leaders and entrepreneurs in the area first visited it as tourists before moving there (as did many other residents).[41]

Given these findings, the New West School believes that communities should pay more attention to the role that quality of life and individual preferences about where to live (preferences that might not maximize income) play in regional growth. They argue that indiscriminate job-grubbing, especially if it includes, as it often has in communities across the West, trying to keep a lumber mill or mine open, or even to reopen closed mills and mines, may harm the very amenities that can secure the new economy.[42] Towns that thrive *after* the chief old-economy employer closes down provide evidence for their argument.[43] Communities that flourish after turning down old-economy opportunities are also fodder for their case, such as New Castle, Colorado, now caught up in the resort-zone growth that has spilled downvalley from Aspen, and where town leaders recently fought the reopening of local coal mines.[44]

Ironically, the development proponents that the New West School says are stuck in outdated models also assume that landscape amenities help drive the growth of western cities, but believe they must make the case to employers, not employees. Urban economic development directors everywhere, from Phoenix to Seattle, argue that recreational attractions such as open space, trail systems, ski areas, and mountain and desert scenery are important selling points in their box of lures for relocating corporations (which, of course, also includes urban amenities such as professional sports teams, well-connected airports, and tax breaks).

But employees' locational preferences might also drive corporate strategy, to some extent, via both push and pull factors. A large corporate example sprawls on U.S. Highway 36 between Boulder and Denver. In the late 1990s, Sun Microsystems began developing a 3,000-employee campus in the Interlocken office park near Broomfield. Sun transferred many employees from California's Silicon Valley, where, some of them told me, the cost of living and crowdedness finally tipped the scales into a degraded quality of life. Simultaneously, a start-up company, Level 3 Communications, built a 2,000-employee complex in the same office park. Level 3's CEO reported that he chose the Colorado Front Range location because a national survey of 500 college seniors and current high-tech employees revealed where and how his potential workforce wanted to live. He told the *Denver Post* that the area beat out high-tech enclaves near Boston, San Francisco, and the other usual suspects.[45] Such corporate decisions can have multiplier effects. A local newspaper, profiling newcomers to the state, told the story of one family lured from Maryland by a Level 3 job (becoming four of the 44,614 immigrants to Colorado from other states in 1999).[46] Besides the job, they cited quality of life as a draw, but we also learn in the profile that their move was presaged by a sister's move to Colorado from Texas and a brother's move from Pennsylvania. Then, after the family settled in, the husband's retired parents arrived from Florida. Our own preferences, and our ties to family and friends, help drive growth.

The story of Level 3 and its in-migrating employees casts light on the complex mix of individual preferences and micro- and macroeconomic forces engendering western growth. People do care about economics. But people also care about family. And, as any real estate broker knows, people are not only pulled to places, but may also be pushed out of places. A study by Joint Venture: Silicon Valley Network, a regional business/labor organization, cited growth as a factor in high-tech companies leaving Silicon Valley. Joint Venture interviewed 100 company executives and found that housing prices, traffic congestion, and long commutes were perceived as reducing their pick of the workforce.[47]

Planning professor Timothy Duane concludes that although both push and pull factors are at work, push factors—mostly unhappiness with urban conditions that yields "affluent flight" from California's cities—have outweighed pull factors in driving exurban growth in the Sierra Nevada foothills.[48] This conclusion matches those of rural demographers who believe that the growth of small towns in the 1980s and 1990s was fueled by a flight (mostly by middle- to upper-income classes) away from problematic cities and suburbs.[49] But I think that the balance of push and pull factors is difficult to discern, and certainly Americans have been attracted to the open territory outside of cities for decades.

Multiple Forces Another approach to the chicken-or-egg problems in regional development theory is to assume that any phenomenon this complicated inevitably encompasses multiple factors that interact, sometimes driving, sometimes dragging on, one another. There is some evidence that, for at least part of the booming 1990s, job growth did lead population growth. Through the mid-1990s, while the Interior West states grew almost twice as fast as the country as a whole, they also had half or less of the national unemployment rate (2.5–2.7 in states such as Colorado and Arizona, places to which economists assign a "base" rate of unemployment of approximately 5 percent). This suggested that job growth actually outpaced in-migration for a while. Job growth slowed in 2001, as the national economy cooled, and could not have continued for very long anyway, as firms adjusted expansion or relocation decisions to the labor supply. Nevertheless, this job growth acted as a key driver of development, especially around places such as Phoenix, Boise, and Las Vegas, and even in smaller towns, such as Bend and Bozeman, that gained significant high-tech and service jobs.[50]

Geographers Beyers and Nelson argue that employment opportunities flourished disproportionately in many smaller towns in the West, a process that fed on, and further drove, their attractiveness, which had mostly relied on their small-town atmosphere and proximity to public lands.[51] In many respects, this is an old story: jobs, population, and development feed on one another in a sequence of regional growth supported by economic logic (e.g., firms searching for lower operating costs and, simultaneously, suitable labor pools) and encouraged and enabled by state and local governments' pro-growth policies.[52] Micro-level factors also interact with these larger variables; in an increasingly footloose economy, people are more able to act on their personal preferences and to live near specific landscape amenities: open spaces, national parks, and water bodies. The main geographic implication of all this: settlement increasingly sprawls out onto the West's open landscapes.

The specific ways in which additional people, jobs, and economic activity shape patterns of land development depend on a variety of things: individual wealth and residential choices and the commercial, industrial, and infrastructural land uses that accompany a growing economy. For example, as people buy larger homes, homes on larger lots, and second homes, the effects of growth on the landscape increase relative to population, and subdivisions necessarily extrude farther into open lands. The 2000 census revealed that the number of households in the United States is growing faster than the population (a result of complex family dynamics) and that a greater proportion of households are buying homes than ever before. More and larger homes per capita means that the per capita footprint on the western landscape expands.

Enabling Forces

Factors that enable regional growth, such as technological innovations, government subsidies, and public investments, also configure development patterns. Enabling forces explain the "how" of development. How did the driest part of the country develop cities in the desert with green-lawned subdivisions, golf courses, and residential enclaves with fountained pools at their front gates? Why doesn't the increasing distance between city and rural residence deter exurbanization?

Improvements in highway, water, and telecommunications engineering (to name only a few) and technical innovations from air conditioning to four-wheel drive sport-utility vehicles have all enabled settlement of western places previously seen as too hot, too snowy, too dry, or too far away from everywhere else. Much of the enabling of western development comes directly from government: national, state, and local. Government investment in highways, schools, water supply systems, and so forth is sometimes seen as simply accommodating growth and often goes unexamined in public discourse about growth. But such investments are also directly responsible for the patterns we see on the ground. Essentially of one mind—that growth is good—the various levels of government work together, and with business, to promote and enable development, not simply to accommodate it. The "growth machine"[53] or "growth coalition"[54] at work in western cities lobbies for federal investment in water, transportation, and recreation, while luring employers and simultaneously building more roads and schools and sizing facilities, from sewage treatment plants to airports, to absorb more growth. An assemblage of advocates makes money or gains status through growth, and most residents acquiesce, too imbued with an optimistic, American assumption about the goodness of growth to question it or to demand something else. In her study of regional development in the United States, economist Ann Markusen noted that coordinated pro-growth policies on the part of federal, state, and local government are especially strong in the Interior West.[55] Refuting arguments that the region's concentration of federal lands deprives the West of autonomy and development options, her assessment is that, on the whole, the federal government has propelled and enabled western growth, most often with the active cooperation of states, counties, and cities.

State government is the penultimate enabler of growth, fashioning transportation and water systems into a trellis on which growth spreads, and always ready to lure the next company, press the airlines for more connections, and promote the tourism that lures new residents and creates jobs. But the ultimate enablers, the key cogs in the growth machine, are the counties and municipalities, which promote growth as the equivalent of community well-being while participating daily in a web of contradictions: some municipal departments curry the favor of potential new employers while others try desperately to keep up with road, school, water, and sewer construction. This goes on while city councils and county commissions field complaints from residents and echo their concerns about traffic, crowded classrooms, and lost views, but then lobby state and federal representatives to bring home more government investments.[56] Local governments near my home on the Front Range rather awkwardly exhibited this recently: they attempted to bag Boeing's corporate headquarters (which eventually went to Chicago) in the midst of a painful soul-searching over growth. Jim Greer, a local commentator, used his *Denver Post* column to chastise the *Post* and others for hyping the possible Boeing coup while they simultaneously "scolded the legislature for failing to pass growth legislation."[57]

Here again, the forces behind development blur together somewhat, as pro-growth local and state governments can drive as well as enable community growth. But it is worth disentangling them because these two types of forces affect development differently and are best treated differently in our response to growth problems. The enabling role of government, as opposed to its driving role, is pervasive, and often implicit and unquestioned. Upgraded roads and water systems are ostensibly accommodations to past growth rather than enablers of future growth, so even citizens who are concerned about growth may not challenge them. In much of the West, still reeling from a decade of rapid growth, a lot of infrastructure development is passed off as simply "catching up." As the debate over growth becomes more sophisticated, watchdog groups in some communities are successfully pointing out that "neutral" attempts to accommodate growth are not always just that, but also serve to enable future growth. They rightly criticize highway projects offered as solutions to traffic congestion—such as the so-called "Legacy Highway" near Salt Lake City or the new beltway around Denver (both described in chapter 4)—for enabling future sprawl.[58]

Enabling factors can operate in complex, synergistic ways. For example, some of the ski areas in the West occasionally subsidize airline service to mountain airports. The subsidized service, although it may be only seasonal, requires and partly pays for permanent airport upgrades. Rising concern over the safety of mountain landing fields also brings pressure to improve them. The general improvement of air service then makes nearby towns more appealing to businesses and entrepreneurs.

A convergence of forces seems especially to have enabled development to disperse into rural areas in the West (and elsewhere) over the last couple of decades, passing, as it were, some tipping point that allowed people to act on their preferences for rural and small-town settings. Timothy Duane identified several such forces underpinning Sierra Nevada exurbanization, including deconcentration of metropolitan employment and new information and telecommunications technologies. But he also argued for two more subtle enablers:[59]

- *A shift from manufacturing to services*: People can create, perform, and consume services at any location; indeed, services follow people and entail few transportation costs. But even some manufacturing activities are acquiring the qualities of services. Duane points out that in the fastest-growing manufacturing sectors (such as computers and other electronics), declining product weight and bulk allow plants to locate regardless of transportation nodes, often in an exurban zone.

- *Equity gains*: Less well analyzed as an enabler of western growth, although often discussed by residents of rapidly growing areas, is the mobility of equity that many urbanites gained in the 1980s and 1990s. House prices soared and investments grew so that a class of people found themselves achieving a kind of economic independence previously enjoyed only by the truly wealthy. Many rode this wealth out to the small towns, exurbs, and ranchlands of the West, deploying it into big homes on large lots, ranchettes, hobby farms and ranches, and new businesses in places they always dreamed of living (say, near a resort or national park they had visited on vacation).

Add the intraregional highway system, the rise of "edge cities," and other elements of the decentralization of employment from urban cores to suburbs, and would-be exurbanites can trade a 30–40-minute crosstown commute for one of equal time, but on rural highways and roads, between home in a rural setting and an edge-city job. Rural planning and zoning policies, which place few restrictions on low-density subdivision, further enable this next step out from suburbia.

Shaping Forces

Geographic features, both political and physical, shape development in the West to some extent. In the heart of the Rocky Mountains, for example, development is squeezed into valley-bottom strips of private land, bounded by steep slopes and public lands. In the Sierra Nevada foothills and around Yellowstone National Park, exurban swaths of development press against public land boundaries. Fragmented, powerful local governments shape urbanizing zones through annexation including extended "flag pole" annexations that reach out to distant commercial properties. Finally, of course, land use planning shapes development patterns, although in my view it is one of the weaker forces in this typology, and so far, its potential looms far larger than its application. The "look" of development—the building heights, patterns of subdivision, roads, setbacks, and so forth— is indeed shaped by land use regulation, but the geographic footprint—the "where" of development—is shaped more by economic factors, the goals of property owners, and a few physical factors.

Any discussion of the shape of development also brings us back to the unresolved debate over the relative roles of individual preferences and market and policy structures in determining what gets built. Planners seem genuinely divided over this question, especially when pressed to predict how home buyers will respond to infill projects or to dense neo-urban developments at the suburban fringe. Antisprawl activists and New Urbanists argue that one main reason we get sprawling suburbs is that that's what the builders sell. But the American preference for small-town, rural, and low-density suburban living is well documented and, I believe, is an important force behind the patterns we see on the ground. In his classic study of suburban culture, *Crabgrass Frontier: The Suburbanization of the United States*,[60] Kenneth Jackson argued persuasively that suburbanizing Americans were chasing after an enduring pastoral dream (although not completely forgoing urban-style services). Arthur Nelson made a similar case for exurbanites: enabled by transportation and employment flexibility, they acted on what he called the long-standing American desire "for the Jeffersonian rural life-style," moving even beyond the suburban frontier.[61]

A recent survey by the National Association of Home Builders concludes, predictably, that most Americans prefer freestanding homes in suburban and rural settings.[62] Certainly the home builders have a stake in promulgating these attitudes, but taking their survey at face value reaffirms, I think, homeowner preferences, rather succinctly encapsulated, in all their contradictions, by the NAHB press release:

> When asked to rate the importance of 16 aspects of a home and its location, "houses spread out" received the highest rating, with 62 percent of respondents checking important or very important. This was followed by less traffic in neighborhood, 60 percent; lower property taxes, 55 percent; bigger home, 47 percent; bigger lot, 45 percent; less developed area, 40 percent; away from the city, 39 percent; closer to work, 28 percent; closer to public transportation, 13 percent; smaller house, 10 percent; and smaller lot, 9 percent.[63]

Academic studies also show that American preferences still favor the pastoral ideal: a careful review of the literature on residential preference, and of demographic trends, led University of Southern California planners Dowell Myers and Elizabeth Gearin to predict only a small increase in the proportion of homeowners who choose denser, more pedestrian-friendly developments over the next decade.[64]

All in all, I am impressed with the influence preferences have in shaping the patterns of development, and I believe that current patterns strongly reflect those preferences. Just because developers' argument for the role of preferences supports a status quo that supplies them with healthy investment returns, they are not necessarily wrong in claiming that they build what people want. For the most part, they do exactly that.

Preferences are also a critical force behind the spread of rural residential development. People prefer lower density, open terrain, and expansive views, when they can get them, especially new migrants to the American West, many of whom are self-selected lovers of wide-open spaces. Their ability to act on these preferences is enabled by baby-boomer wealth, improved highways, high-speed telecommunications links, and the movement of jobs to the suburban fringe. Some regulations and conventions, such as subdivision laws that push developers to carve each rural parcel into lots of at least 35 acres in many western states, also guide their land-buying actions. I think the rural real estate ads capture the preference quite nicely. A company advertises ranches in Utah with "mountain views, a river and a 2.2 million acre forest." "Each ranch homesite has spectacular views of the Wasatch, Uinta or Timpanogos mountains," and all are "only 45 minutes from the international airport in Salt Lake City."[65] Another company advertises "a secluded collection of neighborhoods where expansive views of the Front Range border vast acres of open space," where "life seems to slow down when you're bordered on three sides by open space."[66] Of course, these preferences may not have anticipated the trials of mixing residential development with fire-prone forests, wildlife migration corridors, and agricultural operations.

Preferences may be shaped by external factors, especially government investment and policy decisions, and it is a tenet of the anti-sprawl literature that a web of government policies encourages sprawl. Federal tax policy and home ownership goals help establish a geography that engenders "drive until you qualify" residential search patterns that place new development and new (often first-time) home buyers on the suburban edge; it is almost as if an invisible centrifugal force spins new development out to the edge, pauperizing the core.[67] "Drive until you qualify" now also means drive until you find better schools, nice parks, fast beltways (with toll lanes that let you speed for a fee), and new malls and office parks.

At the landscape and regional scale, it is reasonable to assume, especially in the West, that terrain, climate, water, and other physical attributes shape development somewhat, although technology has weakened their power, as has people's tolerance—even preference—for isolation, steep slopes, and other site characteristics shunned in the past. Obviously, the shaping role of the physical environment was strongest in early western development. Native Americans and early European Americans could not permanently occupy most of the higher elevations or the driest deserts; these landscapes now dominate the public lands, so neither will they be settled today. Physical geography affected—or determined—the paths of early transportation routes, the sites of the first towns, and locations of the early resorts; the precedent of physical limits on development filters down through history and continues to shape development, even after the physical constraints are overcome by technology.

Still, the physical environment seems to matter less and less. Development now attracts water, rather than vice versa: some towns with lots of water grow slowly, while others, including some of the driest, grow spectacularly. Build it and water will come, grow it and more water will come, in the New West.[68]

Even as modern residential, resort, and commercial development has managed to insinuate itself into places that were physically off-limits in the past, some political boundaries still shape it. The federal lands mark perhaps the most enduring boundary on western development, delineating the limits of sprawl even where economics, terrain, or hazards such as wildfires or avalanches cannot. Ultimately, the federal lands will shape the overall regional development footprint, especially the U.S. Forest Service (USFS) and Bureau of Land Management (BLM) lands that dominate the higher elevations and desert basins

Box 2.1

Extensive Federal and State Lands Shape the Development of Private Lands in the West

Public lands—predominantly federal forest, park, and grazing lands, but also including state lands—dominate the two extremes of western landscapes: the higher-elevation forested and alpine areas and the lower-elevation desert and scrublands. Combined on a single map (fig. 2.8), the public lands exhibit a complex footprint of land that, for the most part, is off-limits to private development. Although the public lands may host roads, mines, energy developments, clear-cuts, lodges, campgrounds, and other facilities, they will not sprout subdivisions, shopping malls, or office parks. Indeed, given the apparent lack of restraints on private land development in the West, the public lands act as one of the few solid constraints on land use. Public and private lands interact: public land policies on wildfires, species, and water affect nearby private lands, and development presses up to public land boundaries, into the forest fringe and ever closer to the most preserved areas, such as wilderness areas and national parks. The public lands will increasingly be seen as valuable open spaces in the rapidly developing West.

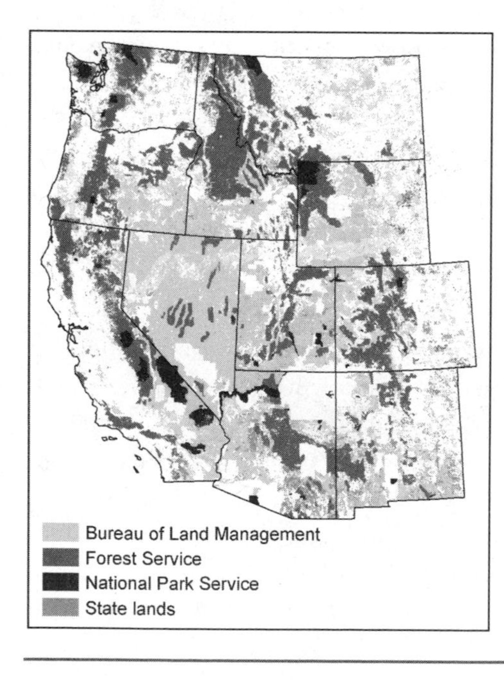

Figure 2.8 Federal and state lands provide a buffer of open space that shapes private land development. All the housing, shopping, office, and industrial developments must fit into the white spaces—the private lands—that intermingle with National Forest, National Park, and Bureau of Land management tracts.

(box 2.1). Particular patterns of federal lands affect private land development. In mountainous areas, for example, private lands are concentrated along valley bottoms and streams; elsewhere, a checkerboard pattern obtains, creating conditions for rural sprawl.[69]

Finally, where does land use planning, the most obvious potential shaping force on western development, fit into this scheme? How effective is it? The land transformations examined in this book are mostly the actions of developers and property owners interacting with local government, which has the power to regulate land use. The processes and outcomes are both idiosyncratic—the result of development decisions made town by town, county by county, parcel by parcel—and universal, as one shopping mall or subdivision looks very much like another. Although Oregon and, to a lesser extent, Washington and California have statewide mandates that shape land development, and even in some cases limit it, land use regulation in the rest of the West is weak, fractured, and uncoordinated. And Oregon's renowned land use planning system was weakened dramatically by an antiplanning referendum in 2004 that favored private property rights.[70] Overall, we cannot speak of an inclusive western land use policy that shapes regional development because each town and county has jurisdiction over land use choices, and because their tools for influencing those choices vary dramatically, as does their eagerness for intervening in the choices of private landowners. I would argue that the region's political-economic regime tends to support whatever landowners wish to do and prefers, in general, land development over land preservation.

Local government has potentially powerful land use zoning and development management authority, routinely upheld by the courts. But a great deal of land use change in the West is occurring outside zoned areas, under the purview of very general county comprehensive plans (if any plans at all) that are only advisory, not compulsory, to planning boards and county commissions. My own reading of dozens of these plans, and my travels through the landscapes to which they apply, suggest that much, and in some cases most, of the development on the ground does not comply with the spirit or the details of comprehensive plans. Even in the rare cases in which a county's rural areas are "zoned"—typically designated agricultural—this is mostly understood to be a holding category until the landowner applies to develop the land.

Few critics have attempted post-audits of planning processes that reveal their actual impact on development patterns. Timothy Duane's blow-by-blow assessment of planning in Nevada County, California, however, is something of a post-audit, and his conclusions support my own sense that government planning is a weak force in the land use universe.[71] I take a closer look at his case study in chapter 5, but, in short, Duane illustrates, in excruciating detail, how plans, once adopted, were not implemented; how earlier actions of various commissions and boards constrained later land use options as conditions changed; how the costs of growth were understated; how buildout numbers and effects were underestimated; and generally, how pro-development interests politically outmaneuvered growth management advocates.[72]

The West also witnessed in the last two decades an insurgency of property rights and antiplanning attitudes and activism, culminating in Oregon's Measure 37.[73] If anything, the net drift in political culture in the West has been toward decreasing the shaping power of government land use planning.[74] This attitude shift enhances the tendency for local decision makers to ignore their own plans. One antidote, increasingly applied across the region, is the lawsuit brought to require adherence to adopted plans. When Gallatin County, Montana, commissioners approved residential and commercial development on Duck Creek, along the western edge of Yellowstone National Park (in an important wildlife migration corridor), the Greater Yellowstone Coalition took the county to court, claiming that the development violated the county's own land use plan for the area. The judge read the relevant plan, and agreed.[75] My experience suggests that literally thousands of developments across the West fall into the same category: they simply do not comply with existing plans, but are permitted because plans are not implemented through binding ordinances, and because watchdog groups cannot afford to fight every bad development.

The power of planning to mold future development in the West is thus up for grabs in a tense interplay between community goals—expressed in plans and backed by planning advocacy groups—and the day-to-day turnings of the growth machine. For the most part, I believe that land use planning has not been a dominant force shaping development, a point further argued in chapter 8.

The Shape of Things

The driving, enabling, and shaping forces behind western land development have evolved over time and will continue to change. Some factors change rather quickly: jobs come and go in a dynamic economy. Others seem more conservative, such as the demographic momentum that will keep the West growing fast and the preference for low-density residential living that drives suburban and exurban sprawl. Some factors seem contradictory. Higher energy prices should be a brake on sprawl, but in the West, where the nation's largest energy reserves lie, higher energy prices mean more economic development and population growth. Still, the diverse set of forces underlying western development provides space for a diverse set of growth management tools to take hold. Enlarging that toolkit requires that we understand the patterns of development in some detail, so part 2 of this book examines each major development landscape more closely. But first, in the next chapter, I take the measure of the current footprint of western development.

3 Footprints of Development

THE DEVELOPMENT GEOGRAPHIES diagnosed in this book are composed of distinct land use patterns, each with a signature pattern of buildings, roads, and other constructions on the landscape. These patterns interact with the unique features of the western landscape to add up to a regional footprint.

In land use terms, "footprint" usually has a rather limited meaning, referring to the total area of a building or other development. In sustainability studies, "footprint" has come to mean very much more: the total of direct and indirect effects of human activity on the global ecosystem.[1] Here I use "footprint" in land use terms, but with a slightly broader meaning to capture the total amount and pattern of land transformed. This footprint can be examined from the regional down to the site scale, parallel to the spectrum of scales used by ecologists, who think in terms of "coarse-filter" to "fine-filter" analysis.

Regional Development Footprint: The Coarse-Filter View

In many ways, the total area of development in the West remains small, especially compared with other regions of the United States. Moreover, a sizeable area of the West will never be developed in most senses of that word. Yet the effect of human activities is pervasive: even wilderness areas are grazed by domestic livestock, thus altering their flora and fauna; dams have changed most of the region's river hydrology; and wildfire suppression has, over the years, shifted forest composition. It is the more visible transformations of land, however, into irrigated agriculture, industry, or residential uses that naturally draw attention in land use studies.

The eleven western states that are the focus of this study comprise some 1.2 million square miles of land (roughly half the size of Australia). Much of this land will never be plowed or subdivided. Slightly more than half of the land in these states is federally owned (in some subregions the proportion runs to 75 percent or more) and thus is not subject to residential and commercial development. State lands encompass quite sizeable swaths of land, too, and they are used in ways similar to federal lands.[2] Additionally, a few local open space programs have grown sufficiently in size to exert regional- or landscape-scale effects on land use patterns. All these public lands, from federal to local, may host roads, trails, utilities, clear-cuts, mines, and a wide range of recreational facilities from ski runs to lodges to restaurants, so it is not accurate to fully dismiss them from an assessment of regional development. The focus here, however, is on the half of the West that is privately owned and potentially subject to development into intensive agricultural, residential, and commercial uses.

Some of the private lands in the West are unlikely to ever be developed. The region's most rugged terrain and arid zones—steep mountainsides and salt-flat playas—even in private hands, may never see more than a few roads, cows, and scattered outbuildings. A few large nonpublic areas, such as Indian reservations, and many large parcels of private lands have seen very little development for various reasons, ranging from physical limits to economic constraints to preferences of the occupants. Increasing areas of private land are covered by formal deed restrictions meant to maintain open space, habitat, and certain uses, such as agriculture, although we do not know the current extent of protected private land in great geographic detail (initial mapping programs are under way in a few states, and regionally, to assess this important, but poorly documented, element of western geography[3]).

We also don't know with great precision how much land in the West is developed or potentially subject to development because of the dearth of reliable land use data in the United States. Only a cursory national database exists, mapped at various times and at very broad scales, or lumped together in county and state totals that are not spatially addressed. Such data are simply inadequate to our need to differentiate among finer land use types at the landscape, or even regional, scale. Detailed land use maps are available for some western cities and counties, and a few states, such as Oregon and California, attempt to map urban and agricultural lands. Probably fewer than half of all the local jurisdictions have complete, current land use inventories, and no attempt has yet been made to combine them into a regional picture.

Still, we can piece together a rough appraisal of regional land use from various sources. The typical starting point here is the Census of Agriculture, one of the few consistent databases on actual land use (although these data are lumped at the county scale, so we cannot determine where in the landscape thc enumerated agricultural uses actually occur[4]). This census is a good starting point because agriculture is the most geographically extensive land use in the West, and because we do know fairly well the extent of intensive agriculture (dryland and irrigated crops, orchards, planted pastures, and feedlots), which works out to roughly 70.3 million square miles (45 million acres), or 20 percent of the eleven western states. Although it is not "development" in the sense employed in most of this book, cropping fundamentally transforms the land (e.g., the natural suite of species is removed, use by wildlife is discouraged, and the hydrology and other physical conditions are altered). The majority of the remaining private open land in the West, and most of the region's public land, is used for livestock grazing, which more subtly transforms the land.

Because the census also accounts for "land in farms" (which is typically more than the land actually cropped or grazed) for each county, its data are often used to track the loss of farmland, a subject of great interest. "Land in farms" can include many different uses, including residential ones, but it is typically employed as a surrogate for agricultural land or, simply, "farmland," as in the phrases "Today, 50 acres of farmland are converted to development every hour," and "More than 50 acrcs of agricultural land are converted to development every day" in the Rocky Mountain region.[5]

Although the numbers don't fully balance, for a variety of data and reporting reasons, essentially all private land use that is not residential, commercial, industrial, or infrastructure is "agricultural" or farmland. It is no surprise then, that less and less farmland shows up in the

Census of Agriculture, nationally and in the West; except in the rare cases of redevelopment, new residential and commercial development is carved from agricultural land, or, in some parts of the West, from timberland or forestland, which shows up in most land accounting systems as agricultural land (fig. 3.1). In accord with the national trend, land in farms has declined in most western states since the early 1970s, although the rate of decline has lessened in recent years (farmland fell 6.2 percent in the 1960s, 2.7 percent in the 1990s, and 1.8 percent between the 1997 and 2002 censuses).[6] Yet "cropland" and, especially, irrigated cropland, the most valuable kind, both of which are more carefully enumerated, show more regional and temporal variation. Total cropland showed a net decline nationally since the 1970s, but fluctuated up and down in the West, actually increasing, from 699 million to 703 million acres, between the 1997 and 2002 censuses. Somehow, cropland appears to be holding on even as "farmland" declines.

Our main interest here is nonagricultural development. The subject has also attracted the interest of the U.S. Department of Agriculture,

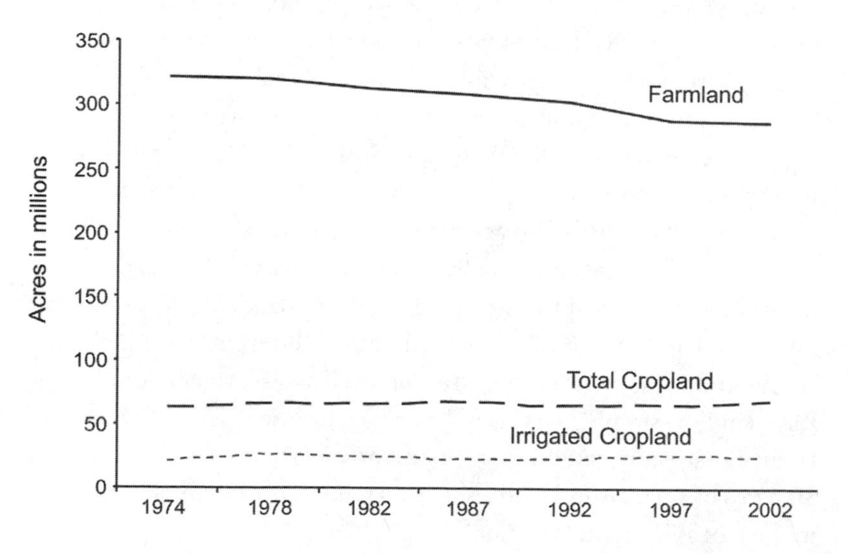

Figure 3.1 Agricultural land trends in the eleven Western states, 1974–2002. Despite a slow net loss of "farmland," actual cropland, both irrigated and dryland, has changed little. This pattern suggests that developed land is carved mostly from agricultural lands not used for crops, including rangeland and abandoned fields. *(U.S. Department of Agriculture, Agricultural Statistics Service.)*

which has been charged, since concerns were first raised in the 1970s about loss of farmland, with tracking the conversion of agricultural land to development. The USDA recognized the limits of its Census of Agriculture in this regard and established another program, the National Resources Inventory (NRI; operated by the Natural Resource Conservation Service, NRCS), to track the status of rural land every five years (since 1982). The NRI was not designed as a detailed land use tracking system; it lumps development into broad categories such as "urban" and "built-up," and its sampling methods have changed over time.[7] But it has become, by default, the only aggregate national assessment of land development in the United States.

The NRI estimated that in 1997, some 15.8 million acres, or 4.5 percent, of the nonfederal land in the eleven western states were "built-up" or urbanized (table 3.1). This area has increased steadily since 1982, when the NRI estimated that 3.2 percent (or 12.7 million acres) were built-up.[8] Nationally, the NRI calculated that approximately 6.7 percent of the nonfederal land in the contiguous United States was developed (the 2002 annual inventory, for which only national data are available, gave 107.3 million acres, or 7.2 percent). The difficulty of gathering such data, as well as problems of data quality, was unwittingly revealed when the NRI had to revise its 1997 estimates of built-up area downward. When originally released, the data were cited as evidence that sprawl had accelerated, but the revision showed development spreading at about the same pace it had for two decades.[9]

The Western Futures project at the University of Colorado's Center of the American West used population data to assess the extent of development (see table 3.1).[10] The Western Futures team mapped the West-wide footprint of development by transforming population into housing counts arrayed in census units. Like the NRI, we first extracted federal land and water, but we also took state land and very steep terrain out of our base of useable land, which left roughly 344 million of the 760 million acres in the eleven western states as "developable" (compared with the NRI's base of 354 million private acres). Even this area includes large swaths with significant development limitations and large tribal lands not likely to develop above rural densities. Of these 344 million theoretically buildable acres, some

Table 3.1

Estimates of developed area for the eleven western states, and as a percentage of private land

USDA/NRI estimates

Year	Urbanized land (acres)	Total land assessed (acres)	% urbanized
1982	12,676,000	398,890,000	3.2
1987	13,772,000	397,960,000	3.5
1992	15,081,000	397,400,000	3.8
1997	15,810,000	354,710,000	4.5

Western Futures estimates

Year	Urban-suburban (acres)	Exurban (acres)	Total land assessed(acres)	% urban-suburban	% urban-exurban
1960	7,687,000	9,389,000	344,269,000	2.2	5.0
1980	12,849,000	13,104,000	344,269,000	3.7	7.5
2000	17,270,000	15,081,000	344,269,000	5.0	9.4
Western Futures projections					
2020	21,719,000	22,763,000	344,269,000	6.9	12.9
2040	27,858,000	28,663,000	344,269,000	8.8	16.4

Note: The USDA Natural Resources Inventory (NRI) uses field sampling to identify "built-up" or "urbanized" area (see: http://www.nrcs.usda.gov/TECHINICAL/NRI/). The Center of the American West's Western Futures project uses population and housing data for census units to map urban, suburban, and exurban development (see http://www.centerwest.org). The two approaches give roughly similar estimates of the development footprint in the West.

17.2 million acres, or 5 percent, were built to urban or suburban densities in 2000. This estimate is quite similar to the NRI's calculation for built-up land in 1997, and perhaps the similarity of the two results lends credence to both. We estimated, however, that another 15.1 million acres in areas that probably did not fall into the NRI's "built-up" category were developed at exurban densities (one unit per 10–40 acres). The remainder was rural, with less than one housing unit per 40 acres.

Together, urban-to-exurban development accounted for some 9.4

percent of the feasibly buildable land in the West. Because another 20 percent of western land is in intensive agricultural use, about a third of the West's private land has been transformed by intensive uses. We projected development to 2040, estimating that urban-to-exurban development would cover some 13 percent of the region's buildable land in 2020, and 16 percent in 2040, growing roughly in proportion to the region's population (see table 3.1). To see how this development plays out on the landscape, we need to know where it is; in other words, we must begin to map its pattern with greater resolution.

Patterns of Development: Fine-Filter Views

Spatial pattern is often neglected in land use analysis, and the aggregate development described above tells us little about the essential geography of western settlement. It is this geography—the actual arrangement of land uses in the western landscape—to which we respond, positively or negatively.

The Satellite View

Unlike those from the Census of Agriculture and NRI, the Western Futures land use data can be mapped to show regional patterns. A "satellite" view (see fig. 2.1) reveals a striking arc of development just inland from the Pacific coast (it adheres to the coast itself in the San Francisco Bay area and in Southern California). This swath, and many lesser ones in the interior, are shaped by topography and by the great swaths of public lands (see fig. 2.8) not subject to private development. A closer look at the footprint of regional development (fig. 3.2) reveals some of the landscapes described in this book, such as the swath of exurban development that fills the Bitterroot Valley and stretches north from Missoula to Kalispell, the Flathead Valley, and the west side of Glacier National Park. To the west is a "twin cities" pattern created by the merging of Spokane, Washington, and Coeur d'Alene, Idaho, with outstretched arms along highways to the northeast and northwest. To the south, in Idaho, a curve of development follows the Snake River from near St. Anthony to Twin Falls. South of that lies the Wasatch Front metro-zone, a slash of development that splits northern Utah.

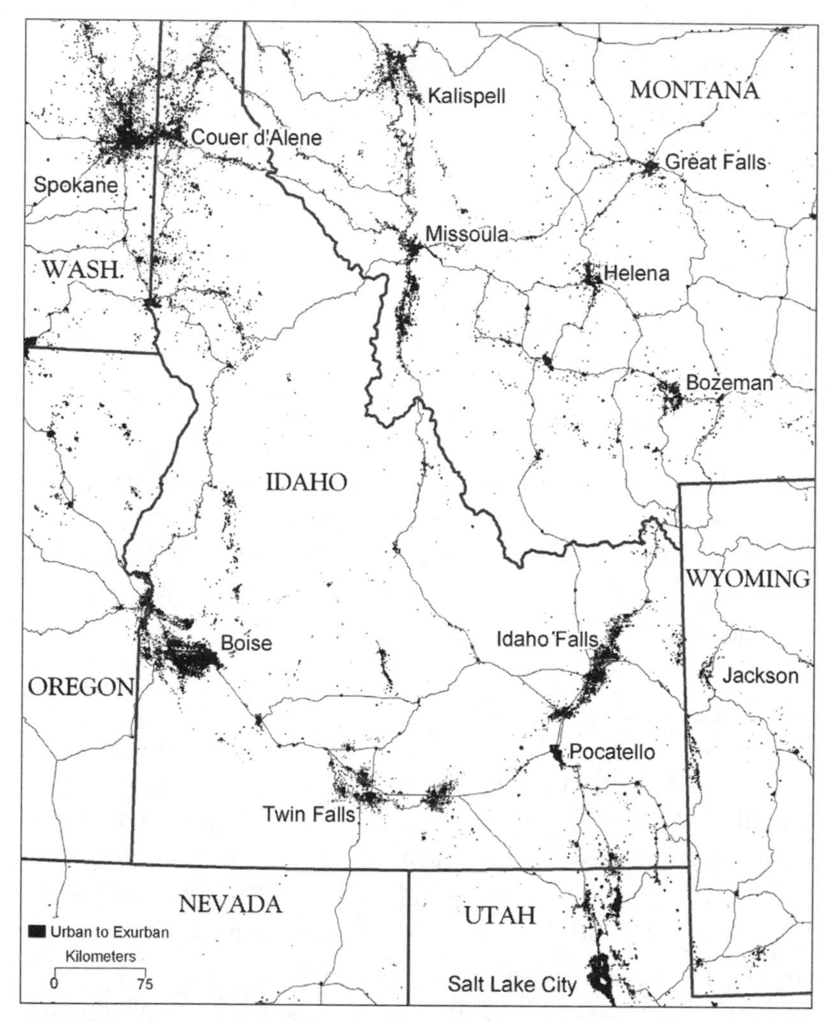

Figure 3.2 A subregional view of the footprint of urban-to-exurban development in the Interior West. Note the many linear patterns, where development follows valley bottoms and is arrayed along mountain fronts.

To better apprehend patterns of human communities and natural habitat, we need to examine the landscape features such as individual cities and large subdivisions that make up the larger swaths of development, as well as the open spaces between towns and the many patterns of natural habitat, such as riparian corridors and forest zones, affected by development. We thus shift from the satellite view to what might be called the "faces" of development.

The Faces of Development

An observer standing on the mountain slopes above Salt Lake City, Boise, or Denver can get a feel for the imprint of those cities on the land: Salt Lake City fills a roughly rectangular valley; Boise elongates along its namesake river; and Denver describes a large half-circle of urbanization, with the flat edge against the foothills and the arc demarcated by the Great Plains steppe that stretches to the east. Ironically, and as the astute westerner knows, such urban areas are actually made more visible by the tree cover they add to otherwise naturally treeless landscapes, although large swaths of pavement and buildings are also visible.

The face of development is more subtle in less urban areas, but nevertheless visible. A hiker above the Bitterroot Valley south of Missoula, Montana, can see, in clear weather, a 40-mile-long, 8–10-mile-wide, flat valley floor with urban forests marking the towns and the widespread stippling of exurban development insinuated into the range, hay, and orchard lands.[11] The valley's exurban development is even more visible when it cuts into the forested slopes (fig. 3.3). In other cases, valley bottoms, where private land dominates, create long zones of agricultural and other developed land uses that are quite striking compared with the less transformed (and often arid) public lands in which they are embedded (plate 1).

The relationships between developed and undeveloped land, and between development and topography, play an important role in shaping sense of place in the American West. Viewsheds are especially important in the West's sparsely vegetated landscape. Planners Christopher J. Duerksen and James van Hemert argue that the West's terrain means that "unlike many other regions of the country where trees and topography can be used to hide development mistakes and bad design, in the West, the often sparse vegetation and sharp landscape features offer little relief."[12] In addition, many western communities lie up against a foothill or mountain backdrop (fig. 3.4). Several western towns, big and small, realized how important this backdrop was when houses began appearing on those slopes, and they scrambled to protect it. Boise, Salt Lake City, and Oakland have all attempted to address development on the nearby mountain slopes that provide their dramatic setting through open space purchases, limits on city services,

Figure 3.3 Roads and homesites in the forested fringe of the Bitterroot Valley, above Hamilton, Montana. Similar exurban development, at densities of one house per acre or less, occurs in many forested settings, wedged between urban and agricultural lands and the federal lands, visible here as the undeveloped swath along the left edge of the aerial photograph. *(USDA Farm Service Agency Aerial Photography Field Office photograph.)*

or hillside development ordinances. I can see the "mountain backdrop" of three different counties (Jefferson, Boulder, and Larimer) from my home-office window, and on a clear day, I can also see large foothill homes 12–14 miles away.

Western land use is so noticeable that a common concern among residents of urbanizing zones is the visual merging of communities as their footprints converge. Lack of landscape separators detracts from community identity, something that city councils, chambers of commerce, and others spend a lot of effort to imbue (with community slogans, fetching logos, festivals, sports teams, etc.). Geographically, sense of place is best achieved by unique layout, notable architecture, and open space separations between one community and the next. Community separators (and "scenic landscape units," areas with scenic qualities and large visibility) are becoming routine parts of master plans, but they remain difficult to safeguard. Private property and development rights assert an owner's ability to build wherever engineering will allow, and liberal annexation laws, as well as differences in development philosophy between cites and counties, make it difficult for any one community to unilaterally protect its geographic setting.

Gradients of Development

A gradient exists around cities, small towns, resorts, and even exurbs, ranging from sites dominated by structures, pavement, and artificial landscaping to development zones where perhaps only 10 percent of the area is so transformed. Geographer David Theobald and others have developed a suite of gradients that capture land use intensity and natural transformations of nature (fig. 3.5).[13] One can trace this urban-to-wildland gradient by traveling a relatively short distance (as few as 20 miles) outward from the core of most western cities. The urban and suburban landscape is fully transformed into buildings,

Figure 3.4 Many western towns derive a sense of place from nearby foothills, and many now seek to limit development in these valuable viewsheds. *(William Travis photograph.)*

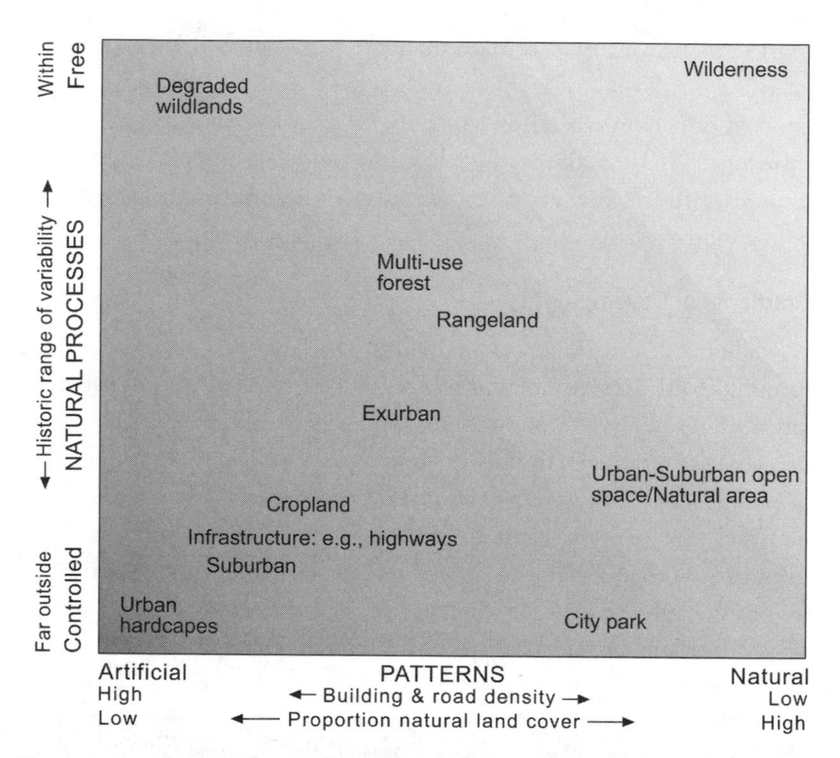

Figure 3.5 The "human modification framework" developed by geographer David Theobald. The vertical axis ranges from areas where natural processes are fully constrained by human development (*bottom*) to areas where those processes are free to affect the landscape (*top*) compared to the historical range of natural variation. The horizontal scale ranges from landscapes fully transformed into artificial, developed surfaces (*left*) to those dominated by natural surfaces (*right*). *(Modified from David M. Theobald, "Placing Exurban Land Use Change in a Human Modification Framework,"* Frontiers in Ecology and Environment *2, no. 3, 2004: 139–144; used with permission of the Ecological Society of America.)*

hardscape, and intensely cultivated landscaping (fig. 3.6). The suburbs come in several different densities and levels of transformation, especially marked by smaller or larger lawns (fig. 3.7). The suburban fringe may be quite sharp, especially where it butts against protected open space (fig. 3.8), or it may exhibit a variety of open and developed spaces interleaved, creating landscapes that can include much land that appears surprisingly natural embedded in landscapes with thousands of homes (plate 2). Exurban landscapes also vary from denser developments that are more like footloose low-density suburbs (fig. 3.9a)

Figure 3.6 Urban hardscape, at the vertex of the "human modification framework," with little or no natural surface and where all natural processes, such as runoff and wind flow, are greatly modified. *(William Travis photograph.)*

Figure 3.7 Three suburban patterns: (a) older, dense subdivision; (b) new subdivision with open space; and (c) the "horse-property," large-lot suburbanization that surrounds many western cities. *(William Travis photograph.)*

Figure 3.8 A typical subdivision plopped down in rangeland near Boulder, Colorado. (above) Roughly 700 homes on approximately 170 acres (four houses per acre), as well as a school and a church, are surrounded by protected open space. *(USDA Farm Service Agency Aerial Photography Field Office photograph.)* (below) The sharp boundary between suburb and open space viewed from the ground. *(William Travis photograph.)*

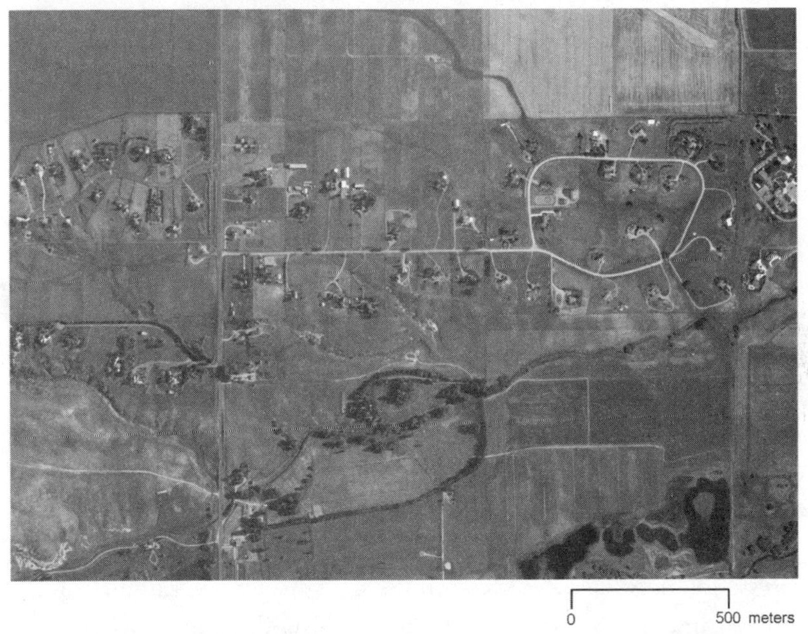

Figure 3.9 The exurban landscape. An exurban development, at about one house per acre, is dominated by houses, driveways, and the modified vegetation of yards and other planted landscapes, along with some seminatural areas. *(USDA Farm Service Agency Aerial Photography Field Office photograph.)*

Figure 3.10 Commercial, industrial, and infrastructure land uses create a growing footprint on the land, especially as nearby populations expand. *(David M. Mixon photograph.)*

to more fully dispersed development (fig. 3.9b), and finally to the isolated home on the range or in the forest. Of course, there's much more to the West's exurban and rural development, including recreational landscapes, which can involve significant transformations of natural cover (plate 3), and all the accoutrements of modern land use, from gravel pits to warehouses (fig. 3.10). Finally, on your transect outward from the city you come to the more traditional rural landscape of farms and ranches.

Beyond the Footprint: Off-Site Effects

In our attention to the edifices of development—housing tracts, shopping malls, and the like—we tend to neglect to account for the land cleared for utilities or ski runs, encompassed in airports, covered by reservoirs, or stripped for building materials. Moreover, we are impressed with what we can see on the land—what the satellite or airborne camera sees—not land use per se. A subdivision, office park, or parking lot comes with lights, noise, stormwater runoff, pets, and other influences that emanate beyond its in situ footprint. Such off-site effects change the local ecology, but are not readily apparent. The Sierra Nevada Ecosystem Project offered a roster of these influences:

> These include habitat conversion and fragmentation; invasion of non-native plants and animals; changes in stream flow and ground water due to land clearing and paving; and increases in ground water extraction, septic effluent and wastewater, fire risk, and fire and fuel-management complexity.[14]

Habitat loss and degradation is a daily reality for ecologists and land use planners. They see each new development, building, driveway, road, water diversion, and wastewater outfall as diminishing the extent and health of natural systems in the West, both those immediately affected and those compromised by off-site effects. The increasing human presence and the cumulative loss of habitat reduce the space available for nature in the West.[15] Around Yellowstone National Park, for example, urban growth and rural residential development have reduced natural habitat and increased the mortality rates of threatened species.[16] In the Pacific Northwest, salmon runs have been decimated by dams, overfishing, logging, and grazing. Now every new road, culvert, and stream of sediment eroding from the individual construction sites associated with both urban and rural sprawl further reduces the species' ability to recover.[17]

Other extensive, although often subtle, off-site effects occur as exurban homeowners demand that nearby wildfires be suppressed or problem bears be removed (or even killed). This enlarging "zone of influence" begins to capture some of the "ecological footprint" concept as it is used in sustainability studies, but we are quite far from being able to make such an accounting of the cumulative effects of land development.

The relationship between on-site development and off-site effects is difficult to pin down. One might reasonably assume that the area of off-site disturbance expands in a manner roughly proportional to primary developments such as housing, retail, office, and industrial. Several factors complicate this relationship, however, and it also changes over time: the number of housing units may grow faster than the population, and the footprint of housing may not be tied to raw population numbers, as people buy second homes and as house size increases. The emergence of "big-box" and regional malls probably also means that commercial space is growing as fast, or faster, than the population, especially in the suburbs. Better indicators are needed, tailored to the different development landscapes.

Patterns of Place

Why is there so much concern about western development, which appears to have progressed perhaps through only a third of the private lands and maybe a fifth of the public lands? One answer is that, although the total footprint matters in many ways, it is the pattern—the geography—of development that affects people's lives. Most westerners live in suburban settings, adjacent to the latest subdivisions and shopping malls. From Bozeman to Santa Fe, they see new development every day, and they notice the open spaces being consumed. Second, it is important to note that development is, to some extent, in the eye of the beholder. Notions such as "developed" or "built-up" remain ambiguous and are certainly subject to various interpretations. Agricultural landscapes, from Great Plains wheatfields to the rice paddies of California's Sacramento Delta, can be seen by casual observers as undeveloped, yet essentially every square inch of such land has been drastically transformed by human action. Even a few houses and other buildings in an otherwise open western landscape give some observers a sense that the area is "developed."

Many westerners identify strongly with the region's open landscapes and wildlands, even while they live in a typical suburb. They appreciate the fact that most of the land remains open, still free of the detritus of urban and rural sprawl, although most may not know whether this is simply a developer's oversight, a private owner's commitment to conservation, or the effect of public ownership. More and more, though, these natural landscapes attract population and economic growth as people try to live as close to open space as their circumstances allow. The region's progressively footloose economy puts more people, houses, roads, and businesses into awkward proximity with remaining natural habitats. The land consumed in this development process and the ecological systems weakened, damaged, and obliterated are essential to regional well-being.

Part Two

Making Sense of the West's Development Landscapes

From edge-city office parks to ritzy ski resorts, swelling land development threatens the ecological integrity of the West and alters the social functioning of its human communities. In many ways, western development emulates national patterns, so we can make some geographic sense of it by examining suburban growth, the emerging exurbs, and the burgeoning resorts. But the region also exhibits unique patterns and problems: its strung-out metro regions encompass large swaths of as yet undeveloped land, and its exurbs and resorts press hard against the last wilderness remaining in the contiguous United States. Even the West's wide-open rangelands are under pressure as new land uses replace traditional ranching. The heart and soul of the West are being whittled away by expanding suburbs, resorts, and ranchettes.

Figure 4.1 The West's urbanized footprint.

4 The Metro-Zones

Shining Cities at the Foot of the Mountains

CITIES AND THEIR SUBURBS take pride of place in most considerations of land use, and although urban areas are more thinly scattered in the West than in the Midwest or East, they still dominate western economics and land uses. Despite the spotlight on rural and agrarian places in most histories of the West, which in their attention to mining, timber, and ranching towns seem to imply that cities didn't exist, a few western historians have brought western cities to the foreground: Carl Abbott recast western history as an urban-centered narrative, and William Robbins analyzed urban-based capitalism as the driver of western history.[1] They pointed out that western cities tended to develop before their agricultural hinterlands, and that urban interests have long called the shots in western development.[2] Authors Peter Wiley and Robert Gottlieb asked, rhetorically: "Isn't it more accurate to see the West as a collection of powerful and expanding metropolitan areas; areas that not only demonstrate the expanding reach of the urban spaces of the West, but also influence, if not dominate, extraurban life and therefore the region as a whole?"[3] Yes, and thus I begin this exploration of the West's evolving land development patterns with a look at its cities (fig. 4.1).

Where Do Cities Come From and What Makes Them Grow?

Despite what many of us learned in grade-school geography, cities don't spontaneously generate from a coalescence of geographic features such as natural ports, travel routes, water sources, and other resource endowments. Although many cities started at a river confluence or important trail junction, others, such as Las Vegas and Tucson, emerged in unlikely geographic settings, and countless promising sites remain unurbanized. Rivers and other transportation routes that followed the grain of topography certainly offered the initial logic for many settlements, and water sources had much to do with the locations of most western cities, but subsequent urban spread followed less constrained features, such as railroads and interstate highways, and city growth was spurred more by political power and the snowballing logic of economies of scale than by geographic endowments. Even a railroad was not an inevitable prescription for urbanization. In the 1870s, the Union Pacific Railroad routed its transcontinental line through what is now Cheyenne, Wyoming, instead of through Denver (which lobbied for it), because the topography west of Cheyenne offered a low-grade path up the first ridge of the Rockies. The decision seemed a blow to the incipient Denver. When surveyor and developer Grenville Dodge laid out the streets of Cheyenne, he predicted that the city would grow to rival, even dwarf, Denver.[4] More than a century later, the railroad still runs through Cheyenne, but the city remains a small northern outpost on the Front Range's 140-mile-long metropolitan corridor, which is anchored, economically and politically, by Denver.

Boosterism and the City

Geography may guide initial settlement, but conscious efforts make cities grow. The real forces behind urban centers are city boosters—urban entrepreneurs—who have goals and a scheme; they grow their cities by getting others—in the West especially, the federal government—to invest in them. Chief among the early western urban boosters was William Gilpin, a politician and real estate promoter who proffered the theory that natural geographic forces—the same ones that created Babylon, Athens, and Paris—were poised to create a great city in the American West, somewhere along what he called the global "axis of intensity" (roughly the thirty-ninth parallel). He first anointed Kansas City as the future midcontinental metropolis, but after moving to Denver, Gilpin concluded that it was the site where urban civilization would blossom in the West.[5] But he didn't rely on harmonic convergence to create development. Instead, he pursued it the way any good developer would: by acquiring and subdividing land, attracting investors, and creating a speculative buzz. Gilpin used his role as a politician to push for growth-enhancing legislation and subsidies. Local leaders and entrepreneurs like Gilpin eventually persuaded the Denver, Northwestern and Pacific (DNW&P) railroad (later the Denver and Rio Grande) to push a line right through the Continental Divide (first over the snow-plagued and unprofitable Rollins Pass, then via the very successful Moffat Tunnel—its motto was "through the Rockies, not around them") so Denver could overcome the Union Pacific's earlier snub.

Boosterism is a key part of the growth process. Phoenix, now the largest city in the Interior West (in both urbanized area and population), was a small, dusty rail town in 1900, no different from dozens of other stops along the Southern Pacific line. A few visionaries, however, believed that Phoenix was poised for growth with its rail service, sunny weather, and a climate said to cure tuberculosis and other ailments. They only needed to get the would-be metropolis over a few development hurdles. Most of all, Phoenix needed a larger, more reliable water supply than the local wells and ephemeral washes could muster. Through aggressive lobbying, local boosters talked Congress in 1906 into placing at the top of the brand new Bureau of Reclamation's project list a dam on the Salt River to collect water for Phoenix.[6] A great irony of the Bureau of Reclamation—created to make the West bloom with irrigated agriculture—is that one of its inaugural projects was an urban water supply system.

Las Vegas, another bright star in the constellation of self-promoting, booming desert cities, started with a tincture of geographic logic, as a way-stop near freshwater springs on the freight and passenger wagon trail between Salt Lake City and Los Angeles. But it grew on unabashed boosterism, helped along by the federal government when the nascent town became a staging area for the construction of Hoover Dam on the Colorado River—another case of the feds seeding and promoting western urbanization.

For the past two decades, Las Vegas has been king of American urban growth.[7] During one of its most notable growth spurts, in the early 1990s, more than six thousand people moved to the Las Vegas area each month. In contrast to many other fast-growing western places, its elected officials simply did not engage in agonizing debates over growth. "We encourage economic development of all types," said the Clark County planning director. "We're trying to accommodate growth rather than limit, control, or cap it in any way."[8] The city offered extremely low taxes, affordability, jobs (often with double-digit rates of job growth over the last two decades), and many amenities. It also ranked high as a retirement town, and retirees constitute perhaps one-fourth of the 100,000 people added to the Las Vegas metro area each year.[9]

Doing their part for boosterism, urban leaders across the West routinely make key development decisions that enhance the growth process. The opening of Denver's new mega-airport in 1995 and Salt Lake City's hosting of the 2002 Winter Olympics certainly were not intended to slow growth. Such boosterism took place simultaneously with debates on growth limits in several western state legislatures in the late 1990s and the appearance of stop-sprawl initiatives on ballots in many western cities.

The Logic of Urban Growth

In addition to boosterism, a self-reinforcing economic logic of urban growth drives the expansion of cities, and has done so for thousands of years, creating Los Angeles–like conurbations from smaller cities. Urban growth is the geographic expression of two economic laws: the economy of scale, in which, past some inflection point, each additional increment of production yields greater return on investment; and the multiplier effect, in which each new job (especially higher-end jobs) or dollar of income creates several support and service jobs and additional income. The sheer scale of suburban-edge developments, in which a single highway exit might be the anchor for ten thousand new homes, allows developers an economy of scale that generates more profits per unit. Even planning departments experience this effect; large planned unit developments (PUDs) require less planning and regulatory oversight per unit than do custom projects, and developers, especially those working at this scale, can be counted on to prepare technically sound plans and to conscientiously adhere to subdivision regulations.

Each new increment of investment also entices, even demands, additional growth (housing construction creates construction jobs, and those workers need houses for their families). Although business economists might argue about how to calculate this multiplier effect for jobs or additional income in a local place (it is typically analyzed at a national level), and although logic dictates that there are some negative feedbacks that retard urban growth, most agree that a multiplier effect occurs, and economic development officers boldly count on it and proclaim it a good thing. Exactly how these forces play out in urban growth is debated by some land use analysts, many of whom worry that the distribution of the benefits of growth is markedly inequitable, especially in terms of taxes and public services, and that this inequity can eventually cause decay and economic retrenchment. The strongest critics argue, simply, that new, low-density suburban growth tends not to cover the costs of the additional services and infrastructure it requires and thus is parasitic on older areas. To cover those costs, city officials redouble their efforts to promote growth, putting their cities on a treadmill and falling further and further behind.[10] In the urban political economy of the United States, no city ever seems able to reach a growth plateau or a sustainable stability.

Once established and growing, cities act as magnets, exerting a socioeconomic gravity that draws money and young people from small towns and rural areas. Cities encompass most of the West's property and sales tax value and host the vast majority of its economic transactions. Cities are, in effect, a form of landscape gravity well (a concept familiar to fans of television science programs in which gravity is typically illustrated as a depression in the space-time fabric surrounding a planet or star into which passing objects inexorably spiral). Indeed, the first geographic models of how cities grew and interacted with one another were actually called "gravity models." Population, equating to mass, could be shown, statistically, to correlate with everything from how many airplanes landed at the airport to the number of businesses offering different services, from rudimentary to high-end. The bigger the city (mass), the bigger the attraction (gravity), and the accumulation of people, money, and resources—in the West, particularly water—would continue. Skeptics who intuitively feel that there must come a point at which this process implodes (and the city becomes the equivalent of a black hole) should consider Los Angeles, which has not yet reached a stopping point.[11]

The Geography of Cities

The fundamental geography of contemporary urban growth in the United States is captured in the fetching, but terribly indefinite, term "sprawl." The term has no widely accepted definition and no universal measure. It generally refers to the low density and large areal extent of American suburban development, but it also adheres to certain features of urban and suburban development, such as the numbing sameness of cookie-cutter subdivisions; big-box, big-parking retail; automobile-dependent development of all sorts; and lifeless edge-city office parks. Because it is not a technical term, planners tend to avoid it in professional analysis, and few have offered a critical threshold of density that can be considered "sprawl." One simple definition of sprawl is urban land use that grows faster than population, which implies that per capita land consumption is increasing, that each new resident's contribution to the urban footprint is growing. Yet, given American urban history and geography, the rate of land development is greater than the rate of population growth in essentially every urban area, even those growing slowly, so this definition doesn't serve us well in distinguishing sprawl from other types of urban growth. Indeed, because their populations grow more slowly but their suburban growth emulates patterns elsewhere, many older, denser cities (such as Boston) appear, by this definition, to be sprawling more than most spread-out western cities, thus calling for more regionally meaningful definitions of sprawl.

Who Sprawls Most?

Antisprawl campaigners like to create lists of the most sprawling cities, but measuring the urban footprint is a lot more difficult than such lists imply. Too many studies rely on assemblages of counties to define urbanized area, despite the inconvenient geographic fact that large areas of many urban counties remain open. Even studies of sprawl that more carefully delineate the urban footprint come with some surprises for those who see the West as especially prone to it. Most assessments of sprawl examine "marginal sprawl," the relationship between recent population growth and additional land consumed by urbanization. Cities everywhere have similar indices of marginal sprawl (approximately an acre of development for every two new residential units), whether the new development adjoins old, dense cities or young, expansive ones, because contemporary suburban development is essentially the same everywhere (except, of course, where effective planning has modified it). But studies of "net sprawl" of whole urban areas, such as the one conducted by the Brookings Institution (which came with its own caveats about data quality similar to those offered above), find that western cities, on the whole, sprawl *less* than their eastern counterparts.[12] The authors, planner William Fulton and colleagues Rolf Pendall, Mai Nguyen, and Alicia Harrison, conclude that northeastern and midwestern cities, while starting out denser in 1982, consumed more land between 1982 and 1997 in urbanization per capita than did southern and western cities. Several older cities grew in area while gaining few new residents, and a few have sprawled while *losing* population.[13] By any definition of sprawl, it is tough to beat a place that spreads out geographically without gaining population! Atlanta was an exception in the Brookings Institution study: it grew in geographic extent more than any other city, but sprawled less (because of significant population gain) than most northeastern and western cities, at least in this analysis.

Fulton and his colleagues suggest that western public lands rein in the spread of the region's cities, and urban analyst Robert Lang has suggested that aridity and steep terrain further limit western urban expansionism.[14] Watching Denver, Salt Lake City, and Phoenix spread out has left me less convinced that public lands or physical limits matter very much; perhaps comparisons such as the Brookings Institution's are simply catching western cities at a particular moment in history when they look more compact than their eastern counterparts.

In his historical study of urban sprawl, Robert Bruegmann uses gross density (total area divided by population) to compare cities. He concludes that most cities worldwide are sprawling in the sense that their density is decreasing. He also finds that the density gradient, a measure of density along a transect from city center to edge, is flattening for most American cities, with the difference between city core and edge decreasing; Bruegmann argues, however, that absolute density is increasing in both downtown and suburban edge.[15]

Debating Sprawl

Different definitions of sprawl play into political arguments about growth management, but these arguments are contradictory and inconsistent. Those who argue against urban growth boundaries and other sprawl-fighting tools like to calculate the average density of Los Angeles so that it exceeds that of Portland, Oregon, which is famous for its growth boundary, mass transit, and planning policies that encourage density.[16] But both cities have been growing for decades, and Portland's growth boundary was not established until 1979. And it works out that Los Angeles exhibits a gross density greater than many American cities.[17] A better analysis would assess recent growth densities, which is what planning professors Yan Song and Gerrit-Jan Knapp did to measure Portland's changing urban form. They found that whereas cities like Los Angeles tend to expand about an acre for every two new dwelling units, Portland adds new housing at densities of roughly eight units per acre and shows increasing concentration, clustering, and transit access.[18]

Nevertheless, Thoreau Institute analyst Randal O'Toole, a critic of Portland planning, cites the apparent contradiction of Los Angeles to support his rejection of most types of urban growth limits: they apparently do not work. But he also sees them as infringements of personal freedoms and abuses of economic efficiency. He argues that Portland's urban growth boundary and pursuit of density was misguided and counterproductive (although, by his own analysis, it appears to have had little actual effect). He is part of a group of growth management critics who have made a cottage industry out of arguing that typical American suburban development is actually desirable, and that anti-sprawl or smart growth efforts are counterproductive.[19]

Arguments over how to measure sprawl can obscure the simple, important fact that western cities do exhibit archetypal American urban sprawl if we simply define it as the spread of suburban land use. It may be slightly denser than metropolitan land use in other regions—what some analysts have begun calling "dense sprawl"—but it transforms large areas of open land nonetheless.[20] Bruegmann concludes that western cities such as Phoenix are exemplars of a flat density gradient; he even attributes to Phoenix "the flattest density gradient in urban history" (although he argues, incorrectly I believe, that many western cities, limited by water and desert terrain, lack an "umbra" of exurbia).[21] Most western cities—like most American cities—also sprawl according to the definition offered earlier: their development footprint increases faster than their population.

Moreover, cities are not islands in rural seas, but more and more are embedded in urbanizing zones including several cities. This means, simply, that any calculation of a city's urbanized footprint that excludes far-flung development that in many functional ways is part of the urbanizing zone greatly understates sprawl. Where to draw the line? The fastest-growing part of the Colorado Front Range urban corridor is in Weld County, 30–50 miles northeast of downtown Denver, in a swath of boomburbs that are not officially part of the Denver Regional Council of Governments. On the Wasatch Front, the fast-growing suburban developments between Ogden and Brigham City are some 45 miles north of the capitol in Salt Lake City.

The debates about which cities sprawl the most, and whether sprawl is good or bad, can also sideline the well-founded concern that western cities are consuming nearby open space and wildlands that provide both cultural and ecological benefits. The West's metro-zones, while accounting for a relatively small proportion of its land area, tend to take over relatively natural habitats, rather than settled rural areas that agriculture and other uses have already heavily transformed. The spread of western cities such as Phoenix and Las Vegas transforms landscapes that still harbor much of their indigenous biodiversity and, often, many endangered species. Edges of Tucson, Reno, Denver, Salt Lake City, and Boise abut national forests or other public lands, often within a stone's throw of designated wilderness or national parks. This juxtaposition of urban and wild makes the West's cities unique and is especially problematic for the conservation of nature in urbanizing regions.

The Geographic Logics of Sprawl

Robert Burchell and his colleagues at the Rutgers University Center for Urban Policy Research point out that suburban-edge, greenfields, and leapfrog development is low-risk investment that appeals to the large development companies: no American developer has yet suffered for building too far out.[22] Indeed, people find themselves attracted to the suburban edge for a variety of reasons, including price, the open space setting, and an abiding American antiurbanism and faith in "newness," which implies better schools, lower crime, and less traffic.[23] Migrants to the urban fringe also attract jobs, schools, and shops nearby, and the cycle of sprawl is complete. Of course, in a region with a continually expanding population and economy, the new, open space–rich suburban developments are inevitably engulfed in the next wave of land development; the only lasting suburban edge is created where the city butts against permanently protected open space. Even in the West, so rich in public lands that cannot urbanize, only a few urban edges are so fixed: growth finds the quadrants of least resistance (north and northeast from Denver; southwest from Boise; south and north from Salt Lake City, and northwest and southwest from Las Vegas) and leapfrogs past blocks of public lands.

Rush to the Edge Businesses discovered that they too benefited from the rush to the edge: companies of all types now settle into low-rise office, manufacturing, and retail parks on the suburban fringe that are close to well-trained, middle-class employees (and, of course, consumers). By 2000, suburbs accounted for about half of all urban office space in the United States, on rough parity with central cities. Suburban office developments sprout low-rise rather than high-rise buildings, thus covering more land per square foot of space.[24] The shift of both jobs and people to the suburban edge continues, with the U.S. Census Bureau reporting, in 2005, that distant suburbs of southern and western cities were the fastest-growing places in America (places the *New York Times*, incorrectly I believe, decided to call "exurbs").[25] With the arrival of jobs and more commercial development, the suburbs grow from bedroom communities to full-blown economic hubs, and they gain tax base. The larger ones attain many of the services and amenities that public administration theory suggested could be efficiently provided only by a core "primate" city (a region's dominant city) acting on its own behalf as well as that of its suburbs.

Annex It The spread of western cities is abetted by land use law. In most western states, suburban cities have powerful rights to annex rural land for urban expansion (Colorado law, for example, allows annexations up to 3 miles from current municipal boundaries). Most cities operate under strong doctrines of local control or home rule, which allow them to annex land without county approval and with little or no notice to nearby municipalities. (Only the landowners' approval is needed, and not that of adjacent landowners.) Cities tend to annex larger swaths than immediately needed to avoid having to fight later with exurban residents who may resist annexation, and to grab tax-rich parcels, such as those by a highway, before another city does.

In the annexation wars that often erupt in growing metro areas, the worst outcome for a city is to be "landlocked"; that is, surrounded by other municipalities. Indeed, urban analyst David Rusk argues that it is better if the central city has the power to, and is situated in a geography that allows it to, annex aggressively so that suburban cities cannot capture the growth and thus fragment the region.[26] Eric Kelly and Barbara Becker recount annexation wars in the Denver and Albuquerque metro areas (the latter is where Rusk, as mayor, got his annexation battle scars) in which suburban cities annex "far more territory than they can serve at the time of annexation simply to prevent other nearby cities from annexing the same land and thus gaining control of its development."[27] Some cities find themselves practicing defensive annexation to prevent development and to maintain open space on their edges. Rusk may be right that giving special annexation powers to the region's dominant city might actually mitigate inefficient sprawl.

The Infrastructure for Sprawl

Highways connecting cities to their hinterlands allow sprawl as residents take advantage of radial mobility to move farther out. For fifty years, the federal interstate highway system, initiated by President Dwight Eisenhower's signing of the Federal Aid Highway Act on June 29, 1956, has been the preeminent enabler of sprawl, not only because the highways radiate out from cities, but also because the goals of speed and capacity that guided interstate highway construction also yielded the "beltway," which created large cis-urban perimeters of access, intrasuburban commutersheds, and prospects for huge swaths of greenfields development. The ways in which highways have reshaped America, especially by allowing people to live farther from town and workplace, have been well documented in, for example, urban historian Owen Gutfreund's recent and compelling telling of the tale: *Twentieth-Century Sprawl: Highways and the Reshaping of the American Landscape*.[28] Kenneth Jackson, a critic of the suburbs that interstate highways helped build, called the highway system "the most important public works project in United States history."[29] The system is mostly complete, except in the West, where new highways and expanded links are still under construction.

Major western highway projects include the widening and improvement of Interstate 90 where it crosses the Cascade Range to link Seattle to the Columbia Basin; Denver's 470 beltway; Utah's Legacy Highway, meant to help tie together the Wasatch Front; improved and extended stretches of California's Highway 99, which threads through the Great Central Valley; and the 101 loop around Phoenix. All these roads will further suburban and exurban sprawl, but all are presented by transportation agencies as simply catching up with transportation needs occasioned by past development. Land use research has firmly demonstrated that highways are enablers, if not outright drivers, of sprawl because they often achieve their goals, at least temporarily: reducing congestion and making the commute to and from far-flung suburbs easier.[30]

Once it was clear that Denver's proposed new airport and associated 470 beltway were sure bets, and refuting the few land use analysts who claim highways do not attract development, investors from around the country quickly made plans for regional malls, office parks, and housing along the beltway, which logically cut through cheap, open farmland and thus created prime conditions for a land boom (fig. 4.2). In 1999, a spokesperson for the DIA Business Partnership, a private group formed to encourage development around the airport, said that developers were eagerly awaiting completion of each 470 interchange: "To have this amount of land adjacent to an international gateway airport is unique."[31] The Denver Regional Council of Governments' planners were quick to add the beltway (which seemed to violate the organization's own proposed growth boundary) into their growth projections: they forecasted in 2000 that the northeastern segment of the region would attract sixty thousand

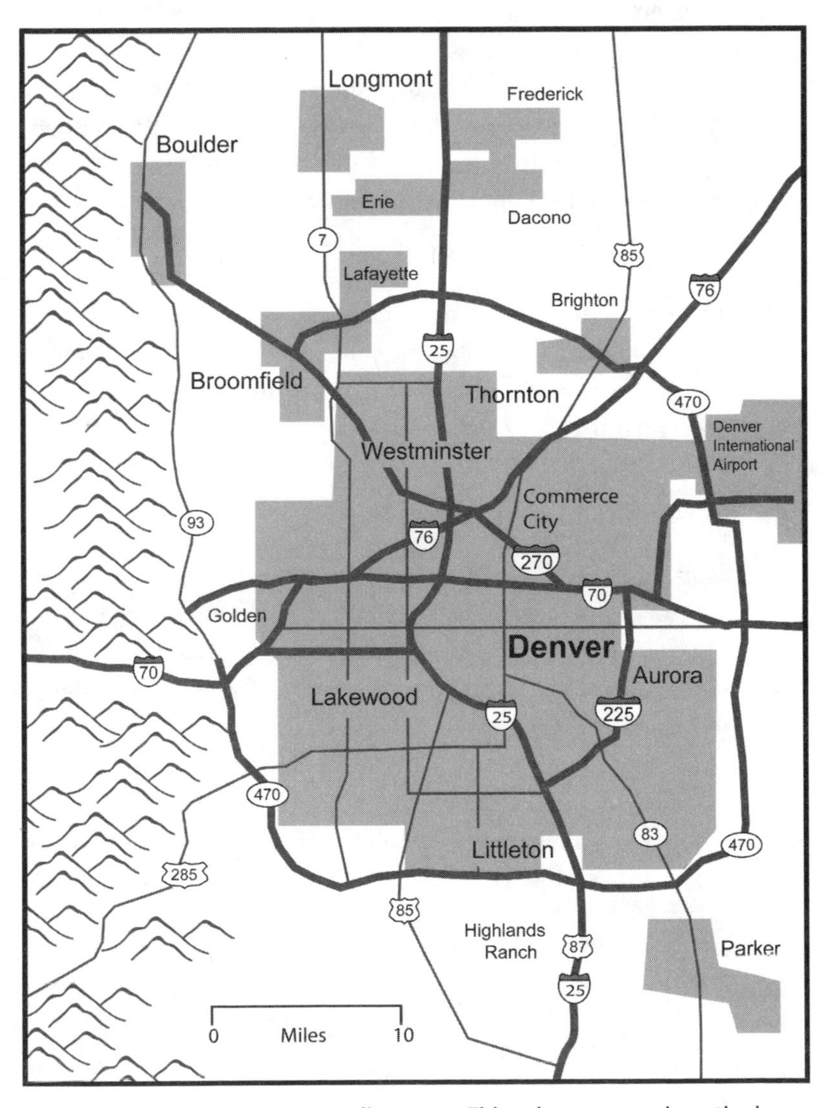

Figure 4.2 The Denver metropolitan area. This urban core anchors the larger Front Range metro-zone. The newly completed beltway (470) extends the urban core's circumference and encourages development on the fringes.

jobs by 2020. By 2003, a leading local planner and architect was predicting that the beltway would generate $35 billion in private investment over twenty years; the DIA Business Partnership thinks $17 billion of this will be along the Denver International Airport sector of the highway. The economic development director for Brighton, a town immediately north of the beltway, noted that for just one interchange, the town had already approved plans for 20 million square feet of commercial development and ten thousand homes.[32] Participants at a DIA Partnership conference entitled "Follow the Yellow Brick Road: E-470 Segment IV. Real Estate's Ruby Slippers?" exulted in the excitement about beltway development. The wicked witch of sprawl did not make an appearance, as far as I can ascertain from the business news coverage of the event; in fact, one presenter was quoted as saying that all this development was not sprawl, "because sprawl is unplanned growth, while all the cities along the 470 beltway have been planning for development for years."[33] But is anyone really planning the West's expanding metro areas?

New Metropolitan Geographies

Urban analysts are rethinking cities—again. Their attention is now increasingly drawn to assemblages of cities in metropolitan regions. Decades after geographer Jean Gottmann popularized the term "megalopolis" to describe the urban complex from Boston to Washington,[34] urban analysts are expanding the concept and applying it to a broader range of city clusters. Urbanists Robert Lang and Dawn Dhavale, at Virginia Tech's Metropolitan Institute, used 2000 census data to identify ten "megapolitan" areas in the United States based on a variety of demographic, economic, and cultural features.[35] Four of these are in the West (fig. 4.3).

In designating new megapolitan regions, Lang and Dhavale argue that "fast growth and massive decentralization transformed once distant cities into galaxies and corridors of linked urban space." This process continued a trend that Gottmann characterized for the Northeast:

> Every city in this region spreads out far and wide around its original nucleus; it grows amidst an irregularly colloidal mixture of

rural and suburban landscapes; it melts on broad fronts with other mixtures, of somewhat similar though different texture, belonging to the suburban neighborhoods of other cities.[36]

Basing their analysis largely on population, Lang and Dhavale find only four megapolitan areas in the West (which they have named Cascadia, NorCal, Southland, and Valley of the Sun), although they suggest that the Colorado Front Range is on the verge. But in terms of

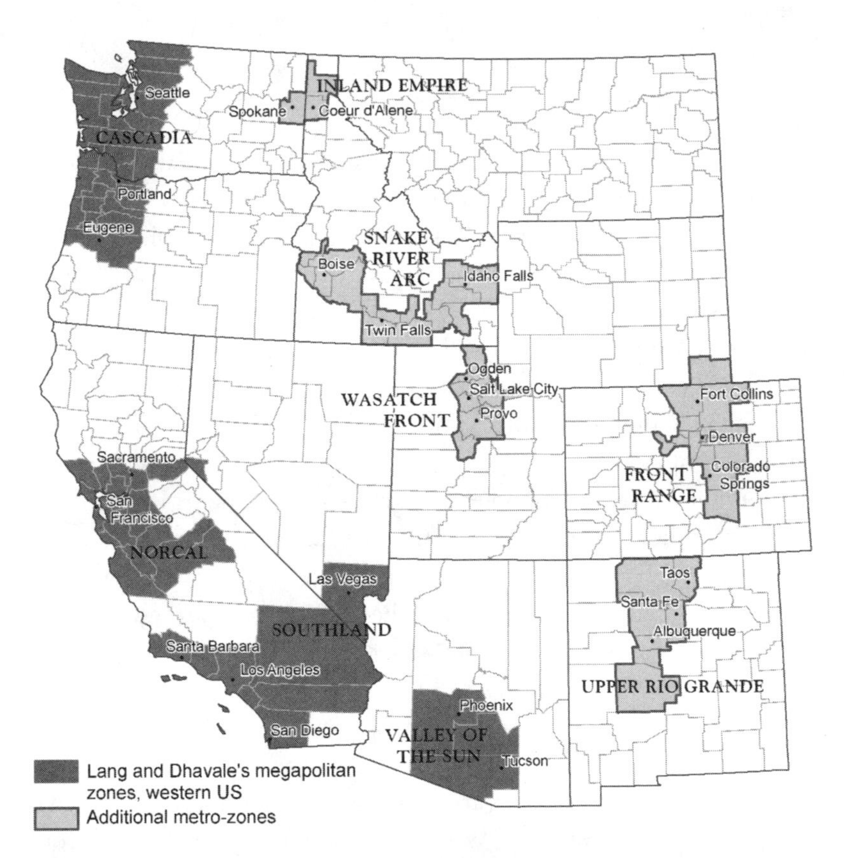

Figure 4.3 "Megapolitan" areas in the eleven western states. Coined by Robert E. Lang and Dawn Dhavale, at the Metropolitan Institute at Virginia Tech, the term "megapolitan area" refers to a large, functionally and economically linked constellation of towns, cities, and whole metropolitan areas. Western megapolitan areas delineated by Lang and Dhavale are shown in dark gray and taken from Robert E. Lang and Dawn Dhavale, *Beyond Megalopolis: Exploring America's New "Megapolitan" Geography*; my proposed additions are light gray. *(Blacksburg: Metropolitan Institute at Virginia Tech, 2005), 13.*

land use (if not demographic) features, it is clear that several other clusters of western cities have many megapolitan characteristics, including not only the Front Range, but also the Wasatch Front, the string of settlements along the Rio Grande from Albuquerque to Taos, and the arc of cities threaded along the Snake River from Caldwell, west of Boise, to Idaho Falls (fig. 4.3). So as not to confuse my approach to these metropolitan geographies with Lang and Dhavale's, I'll use "metro-zone" to refer to these functionally linked assemblages of western cities.[37]

The West's Metro-Zones

The geography of western urban development emulates patterns around the country, but specific regional ingredients create the uniquely western landscape of its metropolitan zones, in which several relatively dense cores anchor a geographic mixture of suburban, rural, and even wild landscapes. The core or primate cities act as government and corporate centers for the metro-zone (although less and less are they key retail centers) and offer regional facilities such as a major airport. Metro-zones grow exponentially as the suburbs of primate cities sprout new commercial nodes; these edge cities then act as anchors for further suburban and exurban development. Urban office space and other facilities move to the increasing circumference of the suburban frontier. Small towns, originally too far from the larger cities to be much affected by growth, find themselves transformed into suburban villages or what some planners are calling "ripple cities": existing towns subsumed in the spread of suburbia and exurbia.[38]

Linear and Galactic Cities

Many western metro-zones are elongated urban corridors aligned with geographic features: the Wasatch Front in Utah, Colorado's Front Range, the arc of the Snake River in southern Idaho, and the axis of California's Central Valley. The likely form of these linear cities in the future has already been set, originally by the topographic features that shaped them, but more recently by investment in the infrastructure that evolved along with them, especially interstate highways. The geometry of linear sprawls gives them a larger footprint than more concentric forms of sprawl. Elongated conurbations offer more edge per unit of urbanized area and more vectors for exurban commuters to intersect the urbanized area; thus they create larger footprints by offering multiple points of job concentration that lend themselves to disparate commuting patterns.[39] The interstices among municipalities attract job centers to greenfield settings within commuting distance of two, three, or more suburban nodes.

The Wasatch Front (fig. 4.4) is squeezed into the narrow isthmus of arid land between the Wasatch Mountains and the Great Salt Lake, paralleling Interstate 15. A drive along I-15, from Brigham City to Nephi, Utah, is a trip through a sprawl of residential subdivisions, commercial strips, and exit-mall clusters that lap against the mountains. The Wasatch Mountains rise dramatically to the east, with snow often covering the highest peaks: Box Elder, Willard, Francis (with its gleaming radar domes), Olympus, Lone Peak, Mount Nebo, and the great wall of Mount Timpanogos looming above Provo.

When I moved to Salt Lake City in the mid-1970s, the city's southerly arc reached only halfway to the Traverse Mountains at the southern end of Salt Lake Valley, and the western slopes were still open land (West Valley City, which now rests on those slopes, did not even exist until 1980 and is now home to over 110,000 residents). In 1975, the ten counties and 100 municipalities that constitute the Wasatch Front urban corridor contained some 800,000 people. Despite the slowed economy in the 1980s, the population swelled to 1.5 million by 1990 and to nearly 2 million by 2000. A combination of immigration from elsewhere and Utah's high fertility rate was growing the population at twice the national rate.[40]

Now development fills the broad Salt Lake Valley. Residents of the "east benches"—prehistoric lakeshores plastered against the mountains that have become prime residential land—can look out across 18 miles of suburban development to the westward-spreading edges of Hunter, West Valley City, and West Jordan. The development continues southward past Provo, insinuating itself into the agricultural villages of Spanish Fork and Payson, and even into the Juab Valley to Nephi. To the north of Salt Lake City, a few vestigial orchards and irrigated farms interrupt an urban strip extending some 50 miles, past Ogden and approaching Brigham City. As in so many western

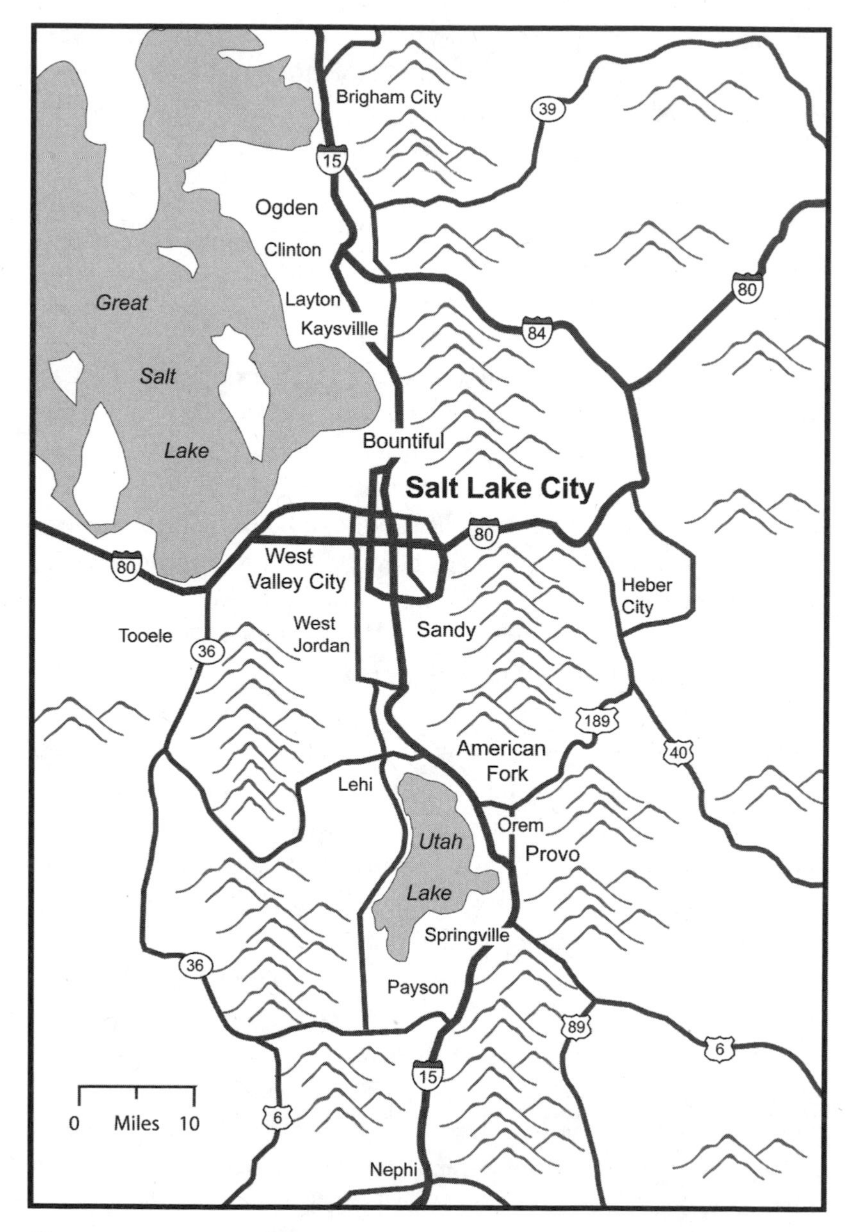

Figure 4.4 The Wasatch Front metro-zone is aligned with the mountains and transportation lines.

linear sprawls, the interstate highway that aided its spread ties it all together.[41]

Anyone driving this stretch could readily understand why Governor Mike Leavitt wanted to squeeze another highway, dubbed the "Legacy Highway," between the interstate and the Great Salt Lake: physical geography demanded that the booming urban swath grow in an elongated north-south corridor.[42] Immediately north of Salt Lake City, the distance between mountain and lake narrows to 2 miles (which must accommodate an interstate and several other roads), and even that choke point was filling with strip development, increasing the pressure to build the highway as soon as possible.

Governor Leavitt had a firm grasp on the area's geographic problem, a problem endemic to the West's expanding metro-zones: to hold the Wasatch Front together as a functional economic entity, he needed to ensure that the dynamic margins of the far-flung urban realm stayed connected to the core, around Salt Lake City. Buyers "driving to qualify" for homes at the southern end of Salt Lake Valley or up to the cheaper housing markets north of Ogden had to be offered a tolerable commute to downtown, the capitol, the airport, the university, and the ski areas if the area were to thrive.

Phoenix offers similar challenges in a more concentric geography. It is sending out galaxy-like appendages in several directions simultaneously (plate 4). In 1998, Arizona State University's Morrison Institute for Public Policy calculated Phoenix's urban area as 9,200 square miles; the metro-zone's circumference was some 150 miles, and the average new home was built 19 miles from the city center, as the crow flies.[43] The largest planned community in the metro area, Douglas Ranch, some 45 miles from downtown Phoenix, in the municipality of Buckeye, will cover 35,000 acres and potentially include 83,000 homes.[44] The new community causes a large bulge in the circumference of the urban area, extending it across the Hassayampa River. Anthem, another planned community, touted as an exemplar of New Urbanism, sits beyond the suburban edge, 32 miles north of downtown. The National Association of Home Builders chose Anthem as "the Best Master Planned Community in America" in 2001. But is this New Urbanism or old-fashioned sprawl? Other major residential and commercial developments extending from the edge of suburban

Phoenix include the incorporated town of Estrella Mountain Ranch, 33 miles to the southwest, and various developments around Queen Creek, 36 miles to the southeast.[45] These multiple, fragmented jurisdictions cause planning nightmares: the editors of the *Arizona Republic* succinctly stated the problem, expressing frustration with "the chaotic mishmash that results from nearly two dozen cities and towns independently plotting their own growth plans."[46]

Metro-Dynamics

Phoenix and the Wasatch Front are good examples of large, functionally integrated but politically fragmented urbanizing areas in the West that have outgrown the authority of local government. The Wasatch Front includes ten counties, some 100 cities, and 160 special districts, all of which aggressively maintain and protect their sovereignty, especially over land use. Many of the cities and other jurisdictions do not think of themselves as "suburbs," or even as part of a larger, urbanizing whole. Their independence, in the face of obvious regional interdependence, makes land use planning especially difficult, despite the patterns of commuting, pollution, and even the visual landscape that perceptibly tie them together.

Suburban cities increasingly challenge the primacy of central cities. The municipalities, counties, and special taxing districts that typify western metro-zones may form a loose confederation, often for very specific purposes, such as regional water or transportation development—or they are sometimes forced to cooperate by federal regulation. In some cases, the suburban cities cooperate with one another in ways that extract power from the old core city (e.g., to fight for a regional airport or to develop their own water supply systems). Many suburbs create significant central business districts (CBDs) of their own (indeed, modern urban planning and design goals often seek exactly this outcome), and economic pauperization of the core city takes hold as firms and retailers move out to the suburbs.

The suburbs find that subsidies and the economic logic of greenfields development allow them to provide urban services, and collect the revenues historically gulped down by the core cities. In the West, this new suburban-city logic applies even to the big engineering efforts needed to get water to residential and industrial customers.

When the Denver Water Board lost its bid to build a big new reservoir as the next component of the area's water supply system in the 1980s, many of its fifteen suburban partners started buying water rights and developing water conveyance facilities of their own.

Metro-Zone Futures

The patterns of future urban expansion in the West seem obvious. Salt Lake City and Denver, guarding, respectively, the western and eastern flanks of the Rocky Mountains, will certainly add to the ends of the transportation axes that already dominate their structure. The "Legacy Highway" in Salt Lake Valley will parallel the existing I-15, allowing the north-south sprawl to lengthen. A private toll road, dubbed the "Super Slab," has been proposed for a swath of Great Plains farmland and rangeland east of and paralleling I-25, uniting the northern and southern ends of the Colorado Front Range metro-zone.[47] Similar infrastructure investments will enlarge other metro-zones. Efforts are under way to improve Highway 99, which provides the backbone for California Central Valley cities that will merge into another linear sprawl. Highway improvements along I-10 southeast of Phoenix will allow sprawl through Casa Grande to Tucson.

Lang and Dhavale already lump Phoenix and Tucson into one megapolitan zone, but when the Western Futures team showed Casa Grande filling in the space between them by 2040 (see plate 4), several locals suggested to us it that would take much longer for that much open space to fill. By 2006, though, when the *Arizona Republic* got a sneak look at a study under way at Arizona State University's Morrison Public Policy Institute, it reported that the two cities were merging faster than experts had predicted only five years ago, with only a 20-mile gap remaining between huge planned developments along Interstate 10.[48] The development has jumped state and tribal lands, so the gap retains some open space, which planners say is a further attraction to development in the zone. Local bets are that the cities will meet at the small town of Eloy, which is already sprouting subdivisions. *Arizona Republic* reporter Catherine Reagor quoted an Eloy resident with an unsurprising attitude: "If I wanted to live in a big city like Mesa or Tucson, I would have moved to one."[49]

5 Beyond the Suburban Frontier

The West's Exurbs

THE TERM "EXURBIA" was coined to describe the scattered, low-density trans-suburban residential settlement around the New York metropolitan area, but exurbanization occurs in near-urban landscapes across the nation, taking a slightly different form in each.[1] On the outskirts of New York City, the low-density residential zone transforms an agricultural landscape that has been settled for generations. In the West, it invades mostly undeveloped, relatively natural landscapes. Rather than cropland and pasture, most of the western land that urban and suburban refugees invade is natural rangeland, which, despite widespread overgrazing, still provides habitat for a wide spectrum of native species, space for natural processes such as fire, a buffer of open space for burgeoning cities, and a tinge of wildness unique to the West.

Exurban development encompasses ten times the amount of land in suburban and urban uses.[2] By one recent estimate, 37 percent of the U.S. population lives in exurban settings, compared with only 8 percent in rural settings and 55 percent in urban areas.[3] In the West, exurbs now encompass more land than do the cities and suburbs themselves. The process of exurbanization expands metropolitan areas as households and businesses seep beyond the urban edge into the rural areas around cities. A study of Portland, Oregon, showed that the charismatic rural areas around the city attracted residents fleeing even the low-density suburbs; some of these exurbanites maintain their city jobs, and others find or make their own economic opportunities in the formerly rural landscape.[4] Westerners are colonizing the mountain and desert hinterlands surrounding cities, creating low-density residential, dispersed commercial, and small-scale industrial developments that scatter the built form of suburbia into the West's wildlands and blur the distinction between urban and rural.

In terms of land use inefficiency and negative effects on natural habitat, exurbanization is the most geographically extensive offender among current development patterns. While smart growth advocates were fighting suburban development, they neglected the spread of large-lot, low-density residential and commercial land uses across large swaths of open land, which may very well turn out to be the most detrimental development landscape in the West. My goal in this chapter is to define this ambiguous development geography—in between the suburban and the rural, a derivative of both, but not truly either—and to assess its social and ecological footprint.

Coining a New Geography

When suburbs emerged in the early 1900s as an important renegotiation of urban geography, we geographers, and other land use analysts, paid some attention, but even then we tossed them in with the matrix of "metropolitan" and left everything else in the realm of the rural. Most land use models draw a line at the suburban edge, calling everything inside urban and everything outside rural—and "rural" was long assumed to be synonymous with agriculture. But a new land use, past the suburban edge but definitely not rural, was emerging on the urban fringe.

To name this settlement past the urban fringe, sociologist A. C. Spectorsky coined the term "exurbanite" (and, by implication, "exurb") in his 1955 book *The Exurbanites.*[5] Spectorsky's focus was on the people who moved beyond the suburbs, not the landscape they created. They were urban refugees who could not abide the conformity of the suburbs, so they simply moved one step farther into the country. Journalists, who often wade in where academics fear to tread, eventually added concepts such as "edge city" and "the exurbs" to the parlance of American development studies. John Tarrant took something of a ruralist's view in his 1976 book *The End of Exurbia: Who Are All These People and Why Do They Want to Ruin Our Town?*[6] Tarrant saw the exurbs as misanthropic and categorized exurbanites as commuting alcoholic husbands, frustrated stay-at-home wives, and bipolar youth; exurbia, for Tarrant, is "the epitome of conspicuous consumption," an elitist construction that will be brought down by energy shortages, its inherent inefficiencies, and the class warfare it represents.

More than a decade later, however, the exurbs were still going strong, and still drawing criticism, although some of it was more thoughtful. John Herbers, in his 1986 book *The New Heartland: America's Flight Beyond the Suburbs and How It Is Changing Our Future*,[7] focused on the exurbs' economic and land use logic. He lamented how little we knew about the exurbs, pointing out that this "new development is neither urban, suburban, rural, nor small town."[8] He defined the exurbs as "new population and commercial growth of very low density, lower than the sprawling suburbs that were decried for scattering urban populations."[9] He compares exurbs to the suburbs: "The new development in outlying places, now at least a decade old, is less noticed and little understood. It is frequently complex and amorphous. Its origins and ideology are more difficult to trace than those of the typical suburban community."[10] Herbers recognized the likely endurance of exurbia:

> It should be obvious by now that even though Americans are unpredictable in many ways, a pattern of growth—from city to suburb to exurb—is underway that has not run its course. The spread of the population into new areas of low density is clearly not the passing fad that some believed it to be. Rather, it is an alternative to both the big cities and their massive suburbs, one that a sizeable number of people have chosen. And nothing on the horizon strongly indicates an end to the trend.[11]

Planners tuned in to the exurbs a bit later. Even in 1994, exurbs showed up as "rural subdivisions," or as "rural neighborhoods" in Randall Arendt's now classic book *Rural by Design*.[12] Exurban development tended to fall through the cracks of professional planning practice and to stay below the radar of county planning staffs and commissions, except in the few places where rural land use regulations existed and where metropolitan planners began to draw urban growth boundaries (UGBs). Planners were having trouble defining, and therefore prescribing, the edge of suburbia as it blurred into an area that was not suburban, but also not rural. Efforts to set urban growth boundaries, like the famous one around Portland, necessitated better definitions of development. In the early 1990s, the "new 'burbs" began to make it into the planning literature as "exurbs," first via a set of studies of development outside Portland's growth boundary.[13] In the Seattle area, after establishing an initial growth boundary, the Puget Sound Council of Governments spent several years coming up with rural and exurban development guidelines. Only a few months after defining a UGB around Denver, the Denver Regional Council looked out past the line, saw the exurbs, and called for a study of what they referred to as "non-urban development."[14]

Only recently have geographers begun making expeditions into this new settlement geography.[15] Exurbia by now has emerged as a development landscape, and the geographic challenge is to define this amorphous concept and its land use in a way that can be quantified and mapped, and to do so for a development landscape whose "natural habitat" is the data-poor, mostly unmapped, and minimally regulated realm of rural counties.

Exurban Geographies of the New West

Across the West, foothill landscapes, at least those not too far from a metro-zone, are sprouting exurbs. Most notable is the exurban zone layered on the foothills of the Sierra Nevada (plate 5), analyzed in planner Timothy Duane's book *Shaping the Sierra*. The steep slopes of Utah's Wasatch Range leave little room for a foothills fringe of exurbia around Salt Lake City, but low-density residential development is spread across the range's gentler eastern slope, locally known as the Wasatch Back.

An extensive foothill zone above Colorado's Front Range cities invites exurban settlement, especially southwest of Denver, where Highway 285 winds through torturous foothill terrain. This is the highway that writer Allen Best drove (and chronicled in *High Country News*) to see how far exurbia extended. At Bailey, an hour into his bumper-to-bumper expedition during the Friday evening rush, "a big yellow sign gleamed into the night, offering 'Alcohol, Firearms, Tobacco,' as if I'd left the genteel exurbs behind and arrived at the wild frontier. Still, headlights glared in my rearview mirror: Yet more exurbanites, in a hurry to get home." At 67 miles from Denver, Best stopped at a store to ask if there were any exurbanites around. The man he asked said he leaves home at four thirty in the morning to get to his job with an excavating company in the Denver suburb of Commerce City. And he knew folks commuting daily from Fairplay, another 30 miles up the highway.[16] Their reward? A house on a mountainside with breathtaking views, deer and elk in the backyard, and a sense of "wide-open spaces," away from the city.

Getting away from city hassles justifies some of this exurbanization, but exurbanites are less attracted to former wheatfields on the Great Plains east of Denver than they are to what the real estate ads call "mountain living near the city." Denver has plenty of space for exurban development on the plains, but the booming exurbs are in the foothills, in the thin forests that drape the Front Range, adjacent to federal lands, including several wilderness areas. Mountains are one of the great attractions of the West, and the ability to live in the mountains—perhaps on a trout stream and adjoining a national forest or park—while still within a reasonable commute to the office is a significant draw for exurbanites.

Foothills are not the only exurban landscape: refugees from suburbia flock to the sagebrush valleys south of Reno and the deserts outside Tucson and Phoenix as well. Exurbs seem to sprout especially profusely in Arizona, partly because state land use laws allow what are widely referred to as "wildcat subdivisions." Rural land developers and subdividers need not worry about paved roads, sewer lines, storm drainage, or the other niceties of formal suburban development; they simply grade roads and build houses. Residents then form special districts to maintain the roads, provide fire protection, and supply water. One such subdivision, Picture Rocks, some 25 miles northwest of Tucson, hosts over ten thousand residents. Reporter Tony Davis described what attracted them: "Those who live here love walks in the desert, and riding horseback. People are drawn by low-key living, by the lack of neon signs, strip malls, and street lights, and by the price."[17] Although some exurban development is pricey—the 40-acre ranchettes with large homes—much of it is relatively cheap: 5-acre lots in Picture Rocks run as low as $500 an acre. The exurbs are a lot cheaper than the suburbs.

Getting away from the big city is not the only push behind exurbanization. Significant exurbs are growing around many small towns in the West, such as Bend, Oregon, Taos, New Mexico, and Flagstaff, Arizona. In Montana, exurban development has spread southwest from Missoula up the Bitterroot Valley and south and north of Bozeman.

Coming into the Country

Population is leaking from the West's cities into its rural hinterlands as newcomers and people already living in the region move farther from urban areas. These exurbanites take advantage of new residential mobility to seek improved quality of life, which, for the mobile class, seems synonymous with low-density settlement. In the 1980s, USDA geographer John Cromartie noticed that nonmetropolitan counties adjacent to metro areas in the West were growing faster than the rural economy would indicate. Places such as Park County, Colorado, and Summit County, Utah, joined the fastest-growing counties in the nation in the 1990s.[18] Analysts also noticed the exurbs gaining jobs: rural counties near cities added jobs faster (44.5 percent job growth from 1985 to 1995) than metro (26.6 percent) or more deeply rural counties (32.5 percent).[19] Even jobs, however, may not be necessary to exurban growth. Several studies show that various forms of nonearnings income (e.g., dividends, interest, rent, and "transfer payments" such as retirement pensions) make up nearly half of all income in the West's rapidly growing rural counties and constitute the bulk of income growth (along with earnings in the service sector) since the early 1990s.[20]

The rather thin literature on the exurbs suggests that two cohorts of people are fueling exurban growth: the city-tethered and the truly mobile. In many ways, the "adjacent, non-metro" places are now all part of the metropolitan geography of the West; their populations rely on the nearby cities, if not for daily jobs, then at least for urban services such as hub airports, entertainment, venture capital, banking, and universities. So western exurbs can be seen as a trans-suburban commuting zone, with its occupants making at least irregular trips into a medium-sized to large city, although several factors can stretch that "commute" a long way. Residents' willingness to drive, say, an hour through open land instead of 30–40 minutes (the national average) through suburbs and business districts enlarges the reach of the exurbs. Just one hour of driving at over 50 miles per hour delivers the exurbanite from home in a wild landscape to city job. The shift in jobs from CBDs to new commercial/residential nodes on the suburban fringe— such as Golden, Colorado, west of Denver, and the industrial and office parks south of Reno, Nevada—made the move to exurbia even easier. The associated shift of office space to the suburban edge—what urban analyst Robert Lang called "office sprawl"[21]—gave exurban commuters an easy target; they need not even drive through the suburbs to get to their jobs. Colorado state demographer Jim Weskott, commenting on Park County's rapid growth (102 percent in the 1990s, from 7,174 to 14,523), nicely encapsulates the western exurb: "It's perfectly situated between the metro area and the mountains and provides that mountain lifestyle while at the same time fairly good access to the [metropolitan area]."[22] The three maxims of real estate hold: location, location, location.

Exactly how far can the exurbs extend? A recent *Washington Post* article on a distant development asks:

> How far are people willing to drive for the privilege of working in the metropolitan area while living in more affordable housing in a more rustic setting?
>
> A hundred miles, one developer is betting.[23]

Exurbanization almost certainly accounts for much of the increase in average commuting times observed nationally over the last decade. The Texas Transportation Institute's first major analysis of commuting in the 2000s found an odd pattern: commuting times increased more in, for example, mostly rural Vermont than in urbanized Connecticut. In two-thirds of the forty-one states for which they had data, the transportation analysts found commutes increasing the most outside metropolitan areas rather than within them. People were commuting longer distances on relatively rural, uncongested roads.[24] These findings suggest that exurbanites are even willing to *increase* their commuting time if most of the drive is through rural areas.

Of course, some exurbanites don't commute. They may transcend the time and distance of commuting with a suite of strategies, such as keeping an apartment in town, or going to the city only to use the airport to start business trips (as do some residents of the Sierra Nevada foothills[25]), or simply avoiding the city altogether. The growing army of self-employed, work-at-home, retired, and simply wealthy people who choose to settle in the West's rural and wild lands are certainly part of exurbanization, although their stories are poorly known. I think most exurban development in the West will remain tethered to the cities, at least loosely, as even the most "footloose" exurbanites want occasional access to urban services, but they want the city to remain "at arm's length," and that arm keeps getting longer.

A "Thudding Sound"

Exurbanites bring "new rural" politics, an affluent chic, and often, a disdain for the rurality to which they were attracted. Detractors blast western exurbanites for their faux rurality, casting them as urban rubes who don't know an elk from a deer, as babes in the woods when it comes to wildland ecology (especially fire ecology). Western commentator Ed Marston chastised them for not recognizing that grazing, logging, and mining had already destroyed the "pristine" landscapes they now colonize: "We've reduced that landscape to a condition where it's good mainly for ranchettes for ex-urbanites who think desert with small trees and broad washes is 'pristine.'"[26] Marston manages to disparage the land uses and landscape knowledge of both old-timers and newcomers in one succinct analysis.

When she somewhat guiltily admitted, in a *High Country News* essay, to loving her 20-acre ranchette outside of Bozeman, Montana, writer Susan Ewing seemed to be inviting criticism, and she got it. It disturbed her that rural sprawl took up valuable farmland and wildlife habitat, but her "craving for space, the outdoors, the company of wildlife, and the chance to settle directly into the ecosystem (of which I am a part)" was stronger than her regret.[27] Letter-writer Auden Schendler retorted: "I'd submit that her ranchette made a thudding, not a settling, sound" into the ecosystem, and that "if the West is going to get off the sprawl highway, we'll need to break away from the misconception that to be part of 'nature' you have to be surrounded by furry animals."[28] Ewing argued that some of her other lifestyle choices (no kids, limited driving) might allow her 20 acres of the Greater Yellowstone Ecosystem, but she was dismissed as just another environmentalist who talks the talk but then walks to her "cabin in the woods." Like suburbanites, exurbanites get no respect.

Geographers Peter Walker and Louise Fortmann examined the tense political ecology of exurbanization in Nevada County, California, in the Sierra Nevada foothills.[29] (It is testament to the geographic concept that, by 2003, they did not bother to define "exurban" in their article, simply assuming that it ranks with urban, suburban, and rural as an established land use classification.) Their detailed analysis unwraps a more complex notion of what's going on in exurbia: in short, locals—rooted at least philosophically, if not economically, in the county's mining and agricultural traditions—see exurban in-migration and settlement as assaults on community; on correct land use; and, because the exurban flood eventually brought calls for land use controls, on property rights. Irony weaves through the story: exurbanites call for land use restrictions to preserve the habitat and views that their own arrival threatened, and longtime residents stand on a proud tradition of sustainable natural resource production, yet they fight restrictions on subdivision as bitterly as they did earlier efforts to limit grazing or timber cutting.

These stories about conflict and criticism of newcomers are nothing new, especially to situations that create such significant changes in land use and social relations, but the exurban landscape these changes are creating *is* relatively new, and it is still little understood and poorly mapped.

Exurbia as a Western Development Landscape

Local governments focus their limited data collection and mapping efforts on zoned areas in and near cities and towns; beyond that lies a terra incognita of land use. Geographer David Theobald found that "a paucity of data, a lack of clear definitions, and a blurring of land use changes and land cover categories make quantifying land use changes beyond the urban fringe" difficult if not impossible in much of the nation.[30] So Theobald used census data on population and housing, rather than land use per se, to estimate the land area in the United States devoted to exurban development. Nationally, Theobald found 378 million acres of exurban development, more than double the area (154 million acres) developed at urban and suburban densities.

Theobald used a similar approach to discern land use in the eleven western states, revealing 3.6 million urban acres, 13.6 million suburban acres, and 15.1 million acres of exurban land use in 2000.[31] The exurbs, then, are almost as large as the cities and their suburbs.

Landscape Patterns in Exurbia

The basic exurban development patterns are founded on two main influences: regulations on rural land subdivision and enabling infrastructure. Rural density guidelines or regulations are fairly similar across the West; by allowing rural land to be carved into parcels of between 35 and 60 acres without coming under subdivision review regulations (as is the case in most western states), these regulations essentially mandate the scattered nature of exurban development. Early notions of appropriate spacing for water wells and septic systems—essential to rural development—were behind the push for large lots, as was the assumption that denser development would occur near municipalities and be annexed to urban services. Most counties also allow subdivision of rural land one or two parcels at a time either at will or over certain time periods, again without more demanding subdivision design regulations. This process, often called minor subdivision and colloquially known as "family subdivision," was meant to allow agricultural landowners to carve out homesites for their offspring. Over time, the process becomes less minor, and can yield extensive rural residential tracts, with each parcel and development independent from the rest, creating an irregular scattering of residences, driveways, and roads.

Other exurban developments materialize on the landscape as multiple parcels created simultaneously, for example, when a large swatch of a farm or ranch is subdivided for residences all at once (fig. 5.1). Geographer Peter Walker and colleagues reconstructed land ownership in Nevada County, California, and found that in 1957, the median land holding was 550 acres, typical in a ranching and timbering landscape, but by 2001, "the 1957 landscape of a few large parcels has been almost completely replaced countywide by a fragmented landscape of many small parcels,"[32] and the median parcel size had declined to 9 acres. As Timothy Duane observed in the same area of the Sierra Nevada foothills, the landscape is subdivided such that—although not all developed and perhaps not meeting current regulations nor befitting updated comprehensive plans—it cannot be put back together.

The Sierra Nevada foothills exemplify a powerful "law" of land use planning: existing parcelization and existing, adjacent land uses set the pattern of future development.[33] This isn't just a case of limited imagination, but a coalescence of social forces. Local officials are hesitant to treat any landowner's development application differently from previous applications. Furthermore, the extant pattern of land use establishes expectations among owners, a hypothetical right of

Figure 5.1 Exurban development south of Bozeman, Montana, where most vegetation remains undisturbed and roads are more obvious features on the landscape than buildings and homesites. (*William Travis photograph.*)

development that local government is loath to challenge, especially in the absence of regulatory zoning. Duane concluded that "perceptions of vested rights differ markedly from legal findings in many cases, but perceptions dominate local politics."[34]

Among the scattered homes that characterize exurbia, another form of exurban development is showing up across the West: the stand-alone suburban-like subdivision, like those outside Tucson, Arizona. In a resort setting, these are likely to be high-end, gated developments, such as golf communities, with large homes on quarter-acre lots. In less ritzy settings, acre lots are carved out of rural land in unincorporated areas, and homes are built without benefit of paved roads, proper drainage, sidewalks, and so forth. Other exurban developments look, for all intents, just like any other suburban subdivision, except that they are plopped down in the middle of rural landscape.

New York Times reporter Felicity Barringer, who covers the sprawl beat nationally, sees this pattern as widespread: "The clash of cultures that has been an inevitable consequence of suburban sprawl for fifty years has slowly changed its context . . . with the ability of ever-more-distant national homebuilding conglomerates to plant dense modern developments far into the countryside."[35] Duane observes that the populations of stand-alone subdivisions in the Sierra Nevada foothills, some reaching ten thousand residents, rival the size of existing towns.[36]

The second main influence on exurban development patterns is transportation infrastructure. Its effect is nested in scale: interstate and other long-haul highways meant to knit the nation together enable the spread of residential and commercial development along widening corridors into far-flung rural areas (as Interstate 80 enables exurbanization of the "Wasatch Back" east of Salt Lake City). At the mesoscale, we find networks of rural roads originally built for access to and transportation of natural resources now acting as conduits for exurbia. Farm, ranch, mining, and timber roads laid down a transportation network that, in the New West, serves the consumption of landscape as amenity rather than as product. This is especially evident in Duane's Sierra Nevada study areas, where rudimentary roads built for gold mining lace the hills and now provide the lattice for exurbanization. Similar patterns of mining roads occur in the foothills above Denver,

Salt Lake City, and Reno and across parts of Montana. A preexisting road network is not necessary in all cases, however, especially where new road construction is relatively cheap and easy: hastily bladed desert roads serve the exurbs outside Phoenix and Tucson.

The Exurban Footprint

In some respects, exurban development can be seen as "light on the land." A couple of buildings—even if one is a 10,000-square-foot log home—and a driveway on 40 acres transform very little actual land. Any standard subdivision at the suburban edge fundamentally alters essentially every square inch of land cover, whereas exurban developments clear, pave over, or build on as little as 10 percent, and rarely more than 25 percent, of a parcel. In landscape ecology terminology, the archetypal exurb mostly "perforates," rather than completely transforms, habitat. That is, houses and even whole subdivisions are relatively isolated, each creating an individual "zone of influence" in a comparatively natural landscape (fig. 5.2). That is one reason why clustering development, instead of letting it spread across the landscape, yields such large benefits: overlapping the disturbance zones around each exurban residence, even partially, yields significant decreases in disturbed habitat.

Roads, driveways, and power and communication lines connect the structures and add linear disturbed areas to the landscape, which can function both as barriers to wildlife migration and as conduits for invasive species. The diffuse effects of exurbanization are even larger, including the impacts of pets on local fauna (house cats and songbirds don't mix), lights, fences, utility lines, garbage cans, and human activities that disturb wildlife but do not show up on a land use map or aerial photograph.

Exurban settlement may expand to cover quite large areas. Just one popular exurb, centered on the foothill towns of Conifer and Evergreen southwest of Denver, in an area bounded by Interstate 70 to the north, the Arapaho and Pike national forests to the west, and the Denver suburbs to the east, laces some 190 square miles with roads, driveways, and houses (Denver itself—the core city and its main, contiguous suburbs—covers roughly 440 square miles) (fig. 5.3).[37] Home to some forty thousand residents, this exurban swath occupies the

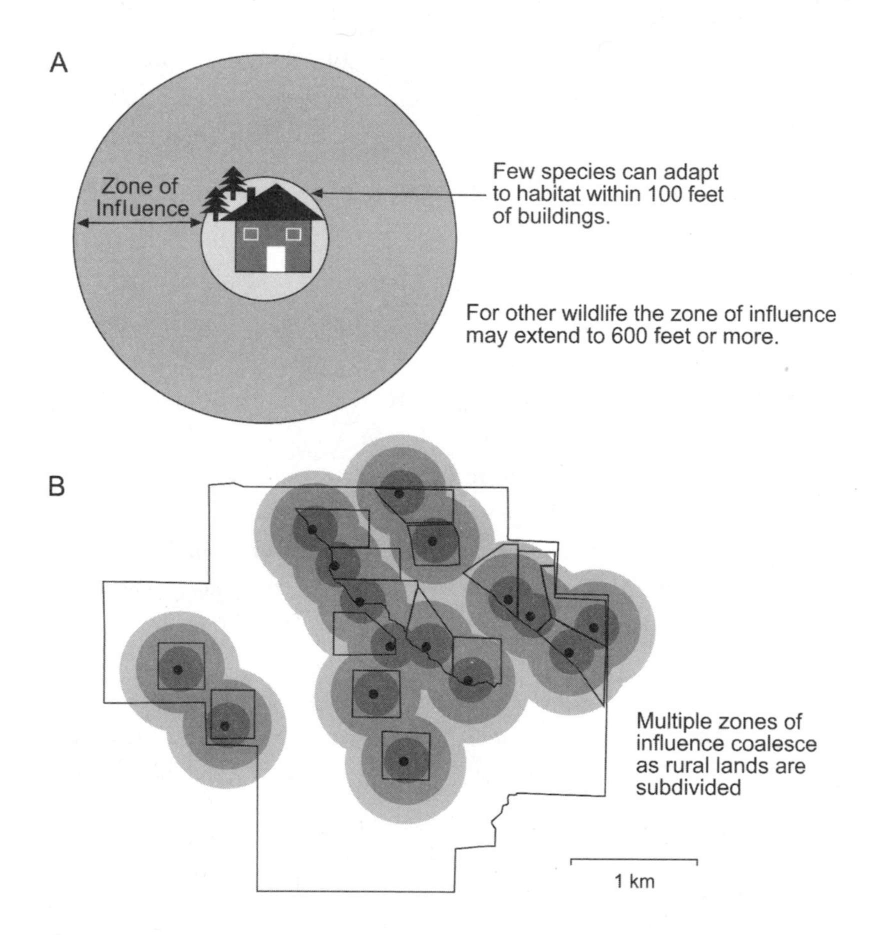

Figure 5.2 Zones of influence around exurban (a) residences and (b) roads. All developments, especially in exurban and rural areas, create disturbance zones where human presence, altered vegetation, noise, lights, fences, pets, sediment, and other elements make the habitat less useable by much of the indigenous wildlife. The extent of these zones varies for different species (ranging up to 600 feet for some species, by some estimates), but the cumulative effect of multiple developments can be reduced by clustering. *(a, modified from Bob Berwyn, "Studies Gauge Sprawl's Impact on Wildlife,"* Denver Post, *May 3, 2003, 1B; and J. D. Maestas, R. L. Knight, and W. C. Gilgert, "Biodiversity Across a Rural Land-Use Gradient,"* Conservation Biology *17, 2003: 1425–1434; b, modified from David Theobald, "Morphology and Effects of Mountain Land Use Change in Colorado: A Multiscale Landscape Analysis," Ph.D. dissertation, Department of Geography, University of Colorado at Boulder, 1995.)*

fire-prone Ponderosa pine stands that characterize the lower montane zone in this part of the Rockies.

Landscape and habitat biases and preferences, like the tendency of Denver exurbanites to occupy the fire-prone forest fringe, exacerbate the effects of low-density development in the West. Land ownership trends, however, also affect the level and type of development effects on the landscape. In most of the mountainous West, the higher ground is federally owned, whereas valley-bottom and riparian habitats are mostly private (and have been ever since homesteaders claimed them in the 1800s). This particular political ecology means, for example, that much of the winter wildlife habitat is private. A study by the University of Wyoming's Institute of Environment and Natural Resources found, for example, that 82 percent of Wyoming's white-tailed deer winter range, and 58 percent of mule deer winter range, is on private land.[38] It also means that everything from houses to roads to railroads to shopping centers must be crammed along streams. A look at the 64,000-square-mile Southern Rockies ecoregion found that roughly

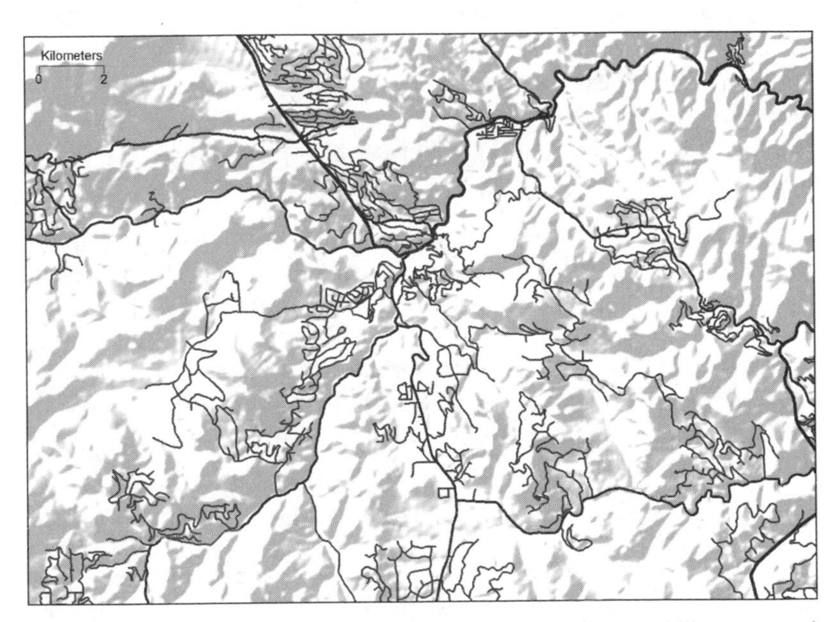

Figure 5.3 An exurban road network in the Front Range foothills. Scattered homesites create many more miles of road per dwelling unit than do even suburban houses.

one-third of the region's entire inventory of streams were within 500 feet of primary or secondary roads (and many more miles of streams were adjacent to dirt roads) and thus were affected by the problems that come with roads, such as polluted runoff, human activity, and invasive species.[39] Although environmentalists may be hesitant to suggest exactly where we should build, surely this riparian bias is problematic in a region naturally short on streamside habitat.

Wildfire The wildfire threat in the West applies to some suburban places (e.g., several areas of California, such as the Oakland Hills, and parts of Santa Fe and Flagstaff, New Mexico, as well as entire towns such as Los Alamos, New Mexico, into the residential heart of which a wildfire roared in 2000[40]). The epicenter of the problem, however, is regional strips where developed and wild land interdigitate: the so-called wildland-urban interface (WUI), or what became known, in the active 2002 fire season, as the "Red Zone."[41]

Newspaper accounts of the 2002 Hayman Fire in Colorado described it as knocking on the door of Denver's suburbs. The fire front was never actually closer than 12 miles to Denver's suburban edge (Highlands Ranch was the suburb closest to the fire), but it did roar through Denver's exurbs. The significant, and growing, wildfire risk in the West is most closely associated with exurban and rural residential development extruding into foothill and forest zones (fig. 5.4). We do not know how much development is at risk, but it almost doesn't matter; as long as homes spread, even thinly, into forest zones, the demand for fire suppression will also spread. As one Forest Service official said (referring to the hills around Montana's Bitterroot Valley):

> We've got too many people moving into heavily wooded areas like this. There's no way we can save a home like this, positioned at the top of a hill, surrounded by all these trees. It's just not going to happen. When people interface with the natural environment in a remote setting, you're going to have problems.[42]

In this way, exurbia is continually diminishing the already limited space for ecosystem processes in the West. Open swaths of federal land on which the public tolerates wildfire are much smaller than maps of national forests and parks would suggest. Meandering bound-

Figure 5.4 An exurban home saved by firefighters, Black Tiger Fire, Boulder County, Colorado, 1986. Development in the "wildland-urban interface" means that firefighters must not only suppress natural fires that might eventually move into the exurban zone, but also expend greater efforts to protect isolated structures. *(William Travis photograph.)*

aries, inholdings, and gooseneck incursions of private land into federal reserves mean that any given acre of federal land is not all that far removed from someone's private land, and probably several residences. In Colorado, where over 75 percent of the forest cover is on public land, fully 80 percent of it is within 2.5 miles of a private land boundary.[43] Each summer in the West, newspaper photographs capture images of worried residents eyeing a huge plume of smoke from a wildfire burning on public lands. As a study by the Greater Yellowstone Coalition concludes: "Because of development occurring on private lands abutting public lands, land managers are being forced to suppress fires that they may otherwise let burn."[44] Fire hazard mitigation makes little difference to the regional outcome. Managers cannot afford to allow fires near "even as residents in the WUI make their homes 'defensible space.'"[45]

Wildlife Real estate ads emphasize the proximity to wildlife that exurbs offer, and indeed, many exurbanites exult in the wildlife they routinely see through their windows. The deer, elk, and coyotes that frequent these homesites add to the "natural," "wild" feel of exurban living, but their presence can be misleading. Although homeowners feel that their effects on the local fauna must be minor (or even positive) because of the abundant wildlife, critics of western development point out that people who build homes in wildlife habitat are destroying the very object of their desire. Dispersed exurban developments appear to compromise a very small proportion of natural habitat, but even in this relatively natural setting, humans soon require wildlife to exist on human terms, so their influence expands dramatically. They are at first enthralled with the wildlife that visits their backyards, but one nasty encounter with elk, for example, can turn wildlife into a problem rather than an amenity. Residents love seeing deer, but not the mountain lions that naturally follow, and prey on, those deer.[46] Sometimes even the deer themselves get to be too much for exurban migrants. *Sacramento Bee* reporter Tom Knudson, in his five-part series on development in the Sierra Nevada, tells the story of Jeff Finn, a California Department of Fish and Game biologist, standing in front of a group of angry homeowners in Lake of the Pines, a foothills subdivision.[47] The homeowners were upset with deer "trampling through yards, gobbling up gardens, even bounding across porches," Finn recounts. "The solution, they said, was to round up the deer and haul them away." Finn, a refreshingly frank public official, goes on: "I told them the problem was theirs, that maybe they shouldn't have put the subdivision there." The two hundred homeowners "wanted my head because we wouldn't do anything." The story becomes less amusing when Finn tells Knudson that every fall, as the deer move down traditional migration routes, "we get calls about deer caught in rope swings, deer injured on fences, deer limping around subdivisions." The deer's migration path through exurbia is killing them.

Mounting scientific evidence points to the significant and widespread negative effects low-density housing has on wildlife, even though most habitat in an exurban development is physically intact. Houses, garbage cans, pets, fences, firebreaks, and even play structures both disturb and attract wildlife, and they set up new territorial tensions between people and wildlife and among wildlife species. Rural subdivisions that become refuges for some animals (e.g., it is widely held that elk know that most exurbs are off-limits to hunting) become death traps for others: bears and mountain lions that follow their prey into these refuges, and act aggressively toward humans, are routinely hunted down and killed. Even species that adapt easily to human presence, such as deer, are compromised in habitats perforated by housing.[48] The concentration of some wildlife, such as deer and elk, into exurban refuges also increases the spread of disease among those populations.[49] Less human-tolerant, often rarer species fare even more poorly at exurban densities. Declines have been documented for the gray fox, several species of birds, and many small mammals in exurban settings.[50]

Cumulative Effects

In consuming habitat and threatening biodiversity, exurbanization is like all other land development patterns. Low-density residential development, however, yields a particularly insidious problem, a version of the "death-by-a-thousand-cuts" syndrome: the ecological effects of each individual exurban development are generally small, yet taken as a whole, they result in significant effects on habitat and species over larger areas. Their cumulative nature can obscure the totality of these effects, making them hard to assess and to mitigate.

The difficulty of discerning and addressing cumulative effects is further exacerbated by the planning process: permitting authorities (typically the county commission) mostly review exurban developments one property at a time, whereas most suburban development comes before planning staff and review boards as whole subdivisions and other planned-unit developments (PUDs). This piecemeal review procedure makes it seem unreasonable to turn down one more home on a piece of land that already holds a dozen or more. The obvious conundrum planning boards confront, as geographer Dave Theobald and colleagues put it, is whether "the future land use of a single property should be restricted because of the cumulative effects of past land use changes on neighboring land."[51] Without considering the ecological effects as they occur incrementally in space and time, such restrictions seem extreme, so rural development often goes unchecked, and individual property owners do not confront restrictions because of likely future effects.

Theobald has worked on this problem for over a decade, employing the obvious tool: maps. The challenge he took on was to illustrate spatially cumulative patterns of development over time, in something close to real time, so that decision makers could see where they were on the curve of, say, aspen habitat loss in their county. They could then assess how each new development would further that curve, providing the objective analytical basis for modifying or rejecting additional developments. Building on the growing power of geographic information systems (GIS) to create complex, dynamic maps, he has illustrated, in ways quite compelling to lay audiences, the spread of development over time and the patterns that result from alternative land use decisions.[52] Tools like this (and others described in chapter 9) provide the means, although not necessarily the will, to assess and to limit the net ecological effect of western development.

Exurbs Unplanned

Whether in a deeply rural county or in the rural part of a metropolitan county, the institutional settings in which exurban development arises offer few land use controls. Exurban areas are typically not zoned, and they are treated as "rural" or "agricultural" land uses in most comprehensive plans, even though these labels no longer apply. Widened highways, liberal well and septic permit systems, and loose building codes all enable exurban sprawl.

Exurbanization not only is an enduring land use trend—and one subject to only minimal planning oversight—but also heightens urban and rural tensions in the West. Eventually, in what demographer Robert Lang called the "triumph of the exurbs,"[53] exurbs become new, stand-alone communities, attracting and creating jobs and other commercial developments, creating their own sprawl and completing the trend toward decentralization in American land use.

6 Resort Geographies

Building a Better Mountain?

WHAT IS THE DIFFERENCE between a "resort" and a "normal" western town? Is Moab, Utah, a long-lived Mormon settlement and mining and ranching town that now attracts mountain bikers, four-wheelers, and plain old tourists, a "resort"? How about Carmel, California? Is Santa Fe, capital of New Mexico, a resort? How about Bozeman, Montana, Boulder, Colorado, or Flagstaff, Arizona? Are these simply normal towns in attractive natural settings getting an injection of the growing tourist trade and the second-home economy in addition to their base economies? Or can we understand their growth and land use patterns better by analyzing them as resorts? The categorization gets easier for places such as Jackson, Telluride, Aspen, Sun Valley, and Vail, the last two built as a resort from the ground up. But not all places that owe much of their local economy to tourism and recreation see themselves as resorts, and many places that think of themselves as based in the workaday economy are actually more in the resort business than some residents and local leaders admit.[1]

The West's resort geography is spreading, drawing more and more communities into its sphere of influence (fig. 6.1). Two themes play out in those communities: first, they rapidly experience several stages of development, each with problems and benefits; and second, the effects of this development, especially on land use, diffuse outward from the core, increasingly affecting a growing rural and small-town landscape.

Figure 6.1 Resort geographies. Places that anchor the resort landscape.

Paradise Paved

One commonality among western places swept up in the resort boom is the surprise and grief that residents express as their towns are "discovered." The rapidity and consequence of community changes overwhelms them. The swift run-up in real estate prices, the second homes built by newcomers that dwarf traditional architecture, new businesses displacing locally owned shops, the growth of service jobs and the service workforce, and a host of other economic changes associated with resort growth have come to be called "Aspenization," after, of course, the place seen as the archetypal resort town, Aspen, Colorado. The global growth of tourism means that more western places will experience Aspenization.

As a member of the Park City, Utah, town council, planner Charles Klingenstein observed these changes firsthand, and he turned to other towns for ideas on what would come next and what might be done about it.[2] Park City is a major destination resort: Deer Valley and Park City Ski Resort offer world-class skiing, and The Canyons, a redeveloping ski area and residential enclave between Park City and I-80, next to the luge and ski jumping complex built for the 2002 Winter Olympics, now offers the most expensive residential real estate in the state of Utah. All this made Summit County, Utah, the fastest-growing county in the United States for much of the 1990s and has kept it in or near the top ten since. Such development, however, is, in a sense, still new to Park City, as journalist Raye Ringholz pointed out in her book *Paradise Paved*.[3] As recently as the mid-1980s, the town's master plan did not anticipate the boom to come. It spoke of the need to lure jobs, retail, and people to boost the city's sluggish economy. The plan specifically predicted the decline of second-home construction, which had progressed in fits and starts.

Only ten years later, however, residents and planners were trying to get a handle on the resort boom; the community's character, and its self-image, had changed dramatically. The population had grown from 2,823 in 1980 to 6,500 in 1996, and it continued growing at 5.4 percent per year. The burgeoning resort real estate market had run into a simple fact of Utah's ski geography: the other well-known areas such as Alta and Snowbird were hemmed in by public lands and narrow canyons. Park City, with its typical mining-era nucleus of Victorian cottages and main street storefronts, was adjacent to large swaths of private ranchlands and, by the early 1990s, had began to sprawl into those open spaces. Big houses, condominiums, shopping malls, and associated infrastructure pushed out into the Snyderville Basin, north along the mountains toward I-80, and across the ridges, over former ranchland thick with scrub oak, east of town. And the downtown was suddenly a marquee of fashionable retail establishments that had more than one Park City resident comparing the place to Aspen.

Besides sending Councilman Klingenstein off to other resorts to gather their lessons in managing growth, and sparking journalist

Ringholz's attention, the speed and magnitude of Park City's transformation nurtured a spokesperson who took on the task of translating the town's experience for other resort residents and planners across the West. Myles Rademan moved to the West in the 1970s, starting as a neighborhood "advocate planner" in Denver and then becoming Crested Butte's town planner. At a time before Crested Butte itself had blossomed into a destination resort and high-end real estate center, Rademan had an inkling of its potential and started preparing the community for what could come. As all the western resorts boomed starting in the late 1980s, many sought out Rademan for advice on how to manage resort growth. Park City snared him as planning director in 1986 because of his style, which was both pragmatic and visionary. He later served as its director of public affairs, with a role in planning for the 2002 Winter Olympics. Having experienced the rapid resortification of two small mining-turned-ski towns, Rademan became known as the "prophet of boom" in resort development circles.[4] This nickname came partly from his hard-nosed message: a big development wave is coming, and you'd better be ready for it. He was sympathetic to the concerns of locals who felt that their towns were changing too fast, and to their pique at, as he says, the "in your face wealth" that inevitably invades resorts, but Rademan would not say what many audiences in booming resorts wanted to hear: that you could fend it off, stay the way you are. Instead, he saw attempts to prevent change in the rural and small-town West as a "fool's dream." The "lifestyle refugees"—a massive wave of well-off baby boomers—were coming no matter what. But even Rademan, having taken on the job of helping other communities get ready for changes bigger than they could imagine, says that "Park City boomed far beyond my anticipation." Certainly Park City's central role in the 2002 Winter Olympics added fuel to the boom, and Rademan has been criticized by some locals for supporting the Olympic bid. He responds that the Olympics would have happened with or without Park City's acquiescence, and he notes that until 1999 there was not even a Park City representative on the Salt Lake Organizing Committee. The trick, he argued, was not to let the Olympics "happen to Park City," but to make sure the town got out of it all it could. After all, as Rademan says in his let's-get-real planning style, "this is what resorts do!"

Resort Development: What Resorts Do

It was the growth of downhill skiing in places such as Sun Valley, Lake Tahoe, and Aspen, especially after World War II, that most defined the nature of the modern western resort and broke trail to the mega-resort.[5] Golf emerged first as a resort amenity adjunct to skiing, then came to dominate desert places essentially dedicated to the sport, such as Palm Springs (which boasts of more than a hundred courses). Other resorts have developed because they are gateways to national parks and wilderness areas (e.g., Mariposa, California; Pagosa Springs, Colorado), are situated in an especially fetching landscape (e.g., Sedona, Arizona; Carmel, California), have great hunting and fishing (e.g., Saratoga, Wyoming), or simply have a particularly attractive mix of cultural and natural features (e.g., Taos, New Mexico). Most resorts have a combination of these qualities.[6]

Much of the early research on the economic and geographic processes associated with tourism, recreation, and the accompanying resort economy was conducted on towns in the Swiss and Austrian Alps.[7] The European literature cites a common roster of effects once resorts are "discovered": employment (especially seasonal employment) and income grow quickly, as do property values, the tax base, habitat disturbance, traffic and air pollution, and of course, demand for services of all sorts. Some of the European studies also tracked the transformation of small resorts, in which most of the businesses were locally owned, into corporate mega-resorts.

All these changes are transforming towns across the West as former ranching and mining outposts are transfigured into high-priced hangouts.[8] Jackson, Wyoming, exemplifies the transition from ranching town to resort, which started rather innocuously in the late 1960s with the development of the two local ski areas and followed a now familiar pattern in which skiing blended into year-round tourism (and souvenir shops replaced the feed and hardware stores). The town grew rapidly as tourism led to real estate investment (and real estate offices and high-end furnishings stores replaced the souvenir shops). Geographer David McGinnis nicely encapsulates this pattern:

> Along with the ski resort came the condominiums and rental units that are standard to world class ski areas. As visitors were

> treated to a slice of life in Jackson Hole, [it] became a contagious place where many . . . were infected with the desire to own a piece of the valley for their own. Soon, real estate developers realized the potential market and began building residential subdivisions along the western edge of the valley. . . . Along came golf courses . . . the requisite "exclusive" tennis club . . . upscale restaurants and specialty shops. . . . While providing the community with a new found cosmopolitan flair, the intended clientele of these establishments was not the locals, but the visitors and wealthy newcomers to the valley.[9]

Jackson's experience fits what analysts call the "resort development cycle." In most formulations, the cycle runs from "discovery" through "rapid development" and ends at "maturity": the mix of businesses stabilizes (dominated by high-end services), visitor numbers fluctuate around a stable mean, and real estate approaches buildout and is owned mostly by nonresidents.

Planner Klingenstein saw nine phases of development in the resorts he studied (Aspen, Sun Valley, Park City, Jackson, and Telluride). One neglected by most resort analysts was a phase of urbanization, in which newcomers are less interested in or tied to the resort base of the community and want more urban services. This leads to phases of conflict and confrontation, followed by the community seeking a new balance among the forces driving its growth. At the time of his work, in 1996, no resort had found the final stage, sustainable balance.[10] Decline or extinction is not an option in most conceptions of resort evolution, although a few studies do recognize the potential for decay of the resort economy, diagnosing it as a form of "self-destruction."[11] A resort that initially offers a unique and attractive experience may simply become too successful at attracting visitors, so that increasing infrastructure and tourist volume inevitably detract from its original attractiveness. Rising real estate prices exclude all but the richest, who themselves may tire of the place, and who squeeze out wage earners so that services are hard to come by. Finally, in its death throes, the "resort sinks under (the) weight of social and environmental problems, [and] most tourists exit—leaving behind derelict tourism facilities."[12] The "locals" then try to revitalize the resort, return to their pre-boom way of life if it is viable, or try something new. I'm not aware of a significant western resort that has entered this phase, although some eastern ones have (e.g., Atlantic City). But such dynamic communities cannot remain stable; their economies change with evolving demographics, preferences, and economic trends, so the future may hold more surprises for resorts.

Resort Economies

Resort development is driven by two basic demands: for recreation and for recreational real estate. There is, of course, a synergy between these two demands, as visitors decide to buy second homes in the resorts where they ski or golf, but there is also some tension. As skier numbers flattened in the 1990s, ski companies got into real estate. Integrating horizontally from ski lifts to restaurants to lodges to vacation homes, they competed with businesses that had plied these trades as secondary to the skiing per se. As resort real estate became more central to further development, full-time and part-time residents came to value the town as a place to live, at least part time, and became less interested in the town's resort qualities and more interested in traditional concerns like public safety, schools, and year-round services.[13]

Resorts do not fit typical economic growth models, and thus economists and demographers have trouble explaining the magnitude of resort development in the West. In the traditional model of economic growth, some basic industry brings dollars into the community and creates jobs, which drive population growth, which then drives the growth of services. Without this so-called "direct basic" income, a community would eventually run out of money as federal and state income taxes slowly drained dollars out of the economy. Direct basic industries replenish tax dollars and fuel growth in the "indirect" sectors, especially services. These two economic sectors provide income for hourly workers and professionals, who spend 30–50 percent of their income on local services, and those dollars then circulate through the community.

It seems obvious that the "direct basic" driver of the economy in resorts is the skiing, golf, or other recreational attractions and, of course, the money that visitors spend pursuing these activities. Yet

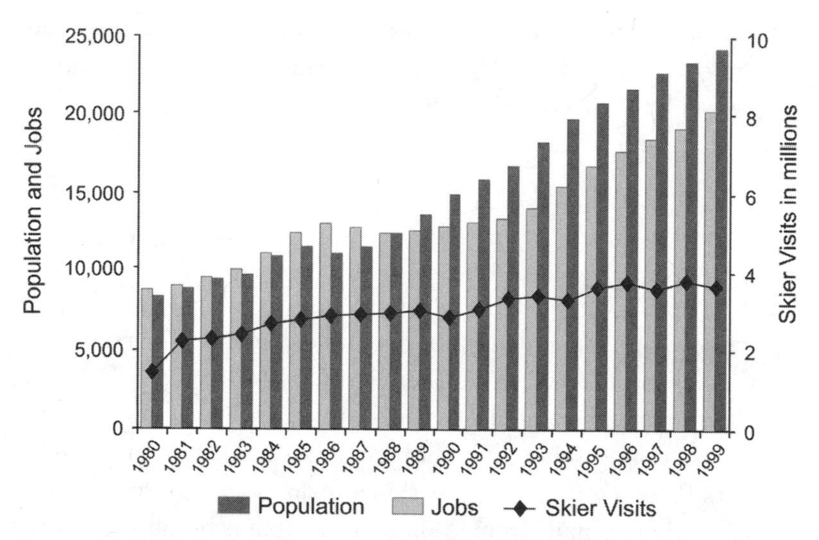

Figure 6.2 Population, jobs, and skier visits, Summit County, Colorado. While skier visits remain stable or grow only slowly, the permanent population and employment in resort counties grows faster, illustrating the maturing of resort economies beyond their original outdoor sports focus. *(Northwest Colorado Council of Governments.)*

many ski town leaders were surprised when skier numbers stopped growing in the 1990s, but their towns still thrived, even grew. Studies of several ski resorts in Colorado found that labor shortages grew even when skier numbers did not (fig. 6.2). Clearly, something other than the destination skier was driving the economy.[14]

As the New West School, championed by economist Thomas Power, has argued, much of the American West by the end of the twentieth century exhibited a different economic equation: amenities attracted people, who invested in their communities and even brought jobs from elsewhere with them, thus fueling the local economy.[15] This equation certainly seemed appropriate for understanding the resorts (although a bit less so for the West's larger cities), and rather than discarding the notion of direct basic industries, some analysts began to redefine a broader array of resort elements as "basic." In particular, construction and the many services required by second homes (even those left empty much of the year) can be seen not as indirect effects of skiing, but rather as primary economic entities that drive the resort economy. One recent study, covering four resort counties in the Colorado Rockies, found that construction and spending on second homes was responsible for a third of all sales taxes, 34 percent of all outside dollars coming into the four-county economy, and 39 percent of all jobs in the four counties (other job sources were winter visitors, 27 percent; summer visitors, 11 percent; resident spending, 16 percent; and other drivers, 7 percent).[16]

Resort employment also follows nontraditional economic principles. The services sector has a high labor demand, and in contrast to many other industries (such as agriculture and manufacturing), there is little opportunity to substitute technology for labor. Therefore, every increase in demand for services must be met by roughly the same increment of increased labor. There are few employment "economies of scale" in tourist and resort residential businesses, so the number of new employees added per unit of business growth remains relatively constant. In high-end resort services—ski services in Vail, golf lessons in Palm Springs, apparel sales in Aspen—the relationship may even be inverted: more employees are needed per customer to provide the desired level of service.

This inverted relationship applies to services for second homes. One hypothetical 10,000-square-foot home in Aspen requires an average of twenty hours of labor per week for maintenance, cleaning, and personal services such as party catering, transportation, and child care.[17] The next such home built will require about the same amount of labor, regardless of the real estate service companies' technical investments. The service providers may find some management efficiencies as they grow, but labor efficiency improvements will always be small, and may even be negative as more and better personal services are demanded (party planning, wrangling, and tutoring). Thus one might surmise that there is a "dis-economy" of scale in high-end resort communities, such that demand for personal services (e.g., number of servers at catered parties, assistants on ski slopes, valet parking attendants, craftsmen, home entertainment technicians, etc.) actually grows per guest, house, or square foot of commercial space. This demand for labor plays out rather awkwardly in the particular geography of western resorts.

Resort Geography

Resorts are important drivers of the West's new economic geography, from the classic ski resorts to out-of-the-way rural retreats. The scale and intensity of development makes resorts stand out in the West's rural landscape. High real estate prices, worker housing problems (in economies rich in relatively low-wage service jobs), and land use idiosyncrasies (such as entire subdivisions of large second homes empty much of the year and upscale retail space more typical of the biggest cities) make resorts a distinct geography with a distinctive cultural and political ecology.

Resort Places

Most western resorts center on distinctive natural amenities, such as hot springs (e.g., Glenwood Springs, Colorado; Desert Hot Springs, California), good hunting and fishing spots (e.g., Livingston, Montana; Saratoga, Wyoming), spectacular mountain or canyon terrain (e.g., Sedona, Arizona; Jackson, Wyoming), and of course, Pacific coast beaches (e.g., Monterey, California). Many combine several attractions, and some are gateways to national parks and other public lands that protect some of those features. Resorts emerged as amenity magnets not long after stagecoaches and railroads first made travel for recreation tolerable in the West. Resort hotels were planned for the hot springs of Yellowstone before reliable wagon roads existed. Dude ranching made Jackson Hole, at the foot of the Tetons, a tourist destination years before Grand Teton National Park was created in 1943, and even longer before downhill skiing made it a destination.[18]

In today's West, even towns not typically considered "resorts," such as St. George, Utah, and Sheridan, Wyoming, are taking on a resort character simply because they lie in attractive settings with pleasant climates. Towns such as Moab, Utah, Santa Fe, New Mexico, and Bozeman, Montana, serve many roles as economic and government centers, but have also taken on the patina of "resorts": as places to visit on vacation, to play on nearby public lands, and even to own a second home. Indeed, the ingredients for resort development are widespread in the West: where is the western town that does not have mountains, deserts, or canyons nearby? As journalist Raye Ringholz observed, "If there's a mountain to hike or ski, redrock backcountry to explore, a waterway to play on, or a desert oasis to green into a golf course, it's being developed by entrepreneurs with hordes of tourists and recreationists hard on their heels."[19]

But the essence of resort development occurs in towns and settings that self-consciously consider themselves to be resorts and shape their economies accordingly. They know quite well what they are—Vail, Sun Valley, Palm Springs—and they make sure that the rest of the world knows they are resorts, that they welcome visitors. If you come and decide to buy a second home, all the better. As Myles Rademan said, "That's what resorts do!" Yet many residents of resort towns are ambivalent about their towns' identity: are they all-out resorts whose main purpose is to serve, and profit from, the itinerant visitor, or are they communities in their own right, needing more attention to the things every town needs and worries about: schools, ball fields, crime, affordable housing, information bandwidth, and well-paying jobs? This ambivalence shows, for instance, in a statement of principle in the Teton County, Wyoming, master plan: "Teton County is a community first, and resort second."[20] But as so many resorts, such as Park City, have found, holding onto community in a resort venue is tough.

The Resort Footprint

Western resort towns are generally small, with permanent populations of less than 10,000, but they host visitor numbers that often eclipse the permanent population by a factor of two or more in some seasons. By definition, a sizeable sector of the local economy relies on those visitors spending money that they earned elsewhere, so most of the commercial space is taken up by lodging, restaurants, and specialized retail. This can be tough on residents wanting more routine services such as non-boutique food and clothing stores, auto repair, and hardware (many find themselves driving to the nearest "normal" town for shopping).

Although generally small, resort towns exhibit larger development footprints than non-resort towns with similar resident populations. Typically, more than half of the residential units in resorts are second homes, whose owners do not show up in census counts. In addition, most second homes are larger than the typical house. Short-term lodging units exceed the number of permanent and second homes. Park City, Utah, for example, with a permanent population of approximately 8,500 (and some 32,000 in the county), hosts up to 20,000 visitors on a peak winter day, with many spending the night. Thus its infrastructure, from lodging to water and sewer lines (not to mention restaurants and bars), must be sized for a community whose population on any given night is more than double its permanent residential population. Other resort towns, such as Aspen, routinely triple their permanent population with nonresident homeowners and visitors. Resorts also include more medium-rise and commercial development than the typical small town. The Park City area includes some 6.5 million square feet of commercial floor space (or 239 square feet per resident), at least twice the commercial space one might expect, or plan for, in a more typical town of its size, or in a suburban area of similar population.

Resorts are scattered across the West, seemingly helter-skelter, their locations dictated mostly by geology and climate, and certainly not by proximity to urban markets. The challenge of resort development is not only to create facilities for visitors in an often wild landscape, but also to provide the transportation infrastructure that brings them to the resort: the early wagon and stagecoach roads into Jackson Hole and Old Faithful, the railroad into Sun Valley, and eventually, the highways and airports that serve Vail, Mammoth Lakes, Palm Springs, and Telluride.

Although most resorts are isolated points of development in otherwise rural areas, some do cluster, creating resort complexes: examples are those on the Monterey peninsula and around Lake Tahoe in California, those in the central Colorado mountains (the largest concentration of ski areas in North America); the beachside towns on the central Oregon Coast, and the desert resorts around Palm Springs. Most western resorts are wrapped in a landscape of public lands; indeed, the naturalness and scenic quality of protected public lands provide the supportive setting for resort development of nearby private lands.[21] At most ski resorts in the West, skiers step onto public lands as they load the lift. But the conventions, golfing, and of course, shopping are all in town, on private, not public, land, as is the key geographic effect of the resort boom: residential real estate development.

Many resorts deploy very high development densities (including high-rise condo and hotel complexes) into rather small physical footprints. Some resorts are mixed in with natural land covers like forest (plate 6).

Several aspects of resorts make their imprint on the land somewhat unique among settlement patterns. First, of course, ski resorts, as mentioned above, involve thousands of acres of ski slopes, including not only deforested ski runs but also seminatural areas between the runs, forest glade skiing areas, dispersed skiing above timberline, and the relatively small footprints of on-slope facilities such as lift and gondola houses and towers and restaurants.

The town of Vail, for example, covers approximately 4.5 square miles (2,900 acres) in an elongated swath along Gore Creek (a good amount of that land is covered by a four-lane interstate highway that runs the length of the valley). The Vail ski area, the largest in the Rockies, includes 8.3 square miles (5,289 acres) of skiable terrain. Other types of resorts also transform landscapes. The hundred or more golf courses around Palm Springs and Palm Desert, California, stand out on aerial photographs even more than ski runs do (turn on the "golf layer" on Google Earth to see this extraordinary assemblage of desert golf greens). Add the "down-valley" effect of development pushed farther and farther out from the core, both for resort uses (such as golf courses at lower elevations in ski towns) and more mundane land uses (worker housing and all the services that can't afford resort real estate, such as landscaping services, big-box retailers, storage units, and construction firms), and the resort footprint expands even more.

The resorts rely on a transportation infrastructure that allows visitors easy access; indeed, they thrive on major transportation infrastructure, laid across rural areas and often across challenging terrain. Highways such as I-70 and I-80 make Vail and Tahoe, respectively, more likely to develop high-end real estate than less well-connected resorts, such as Steamboat Springs or Taos; four-lane stretches of Highways 1 and 101 link Monterey to the San Francisco Bay area. With essentially no public transit available to these otherwise sparsely settled areas, these roads to recreation host weekend and holiday traffic jams not unlike the mass exodus to "the shore" common in the eastern metropolis. Elsewhere, airport access is critical to resorts that cater to wealthy, itinerant homeowners, bringing private jets and large jetliners into small mountain and desert airfields.[22]

The resorts did not create these transportation links, but the largest resorts are big because of them, and highway and airport improvements help grow the resorts. The newest western ski resort (at this writing, at least), Tamarack, and its host town, Donnelly, Idaho, are benefiting from improvements on State Highway 55, a long, winding rural road along the Payette River now called on to carry construction and visitor traffic to the new developments. Resorts across the West have invested their own money, or sought state and federal dollars, to improve local airports, or even to subsidize airline flights into nearby airports.

As in most American communities, however, no transportation improvement is without controversy: citizens fought the widening of State Highway 179 into Sedona, Arizona and residents extracted design concessions from the Wyoming Department of Transportation meant to lessen the effects of enlarging State Road 22 from Jackson through Wilson to the foot of Teton Pass. A few regional highway projects meant to help ease urbanites out to the wildlands and resorts, such as the widening of I-90 from Seattle across the Cascades and the impending decade-long reworking of I-70 from Denver westward across the Rockies, pose even thornier problems for resorts, the towns along those transportation corridors, and of course, natural areas they traverse.[23] Vail homeowners already complain about the noise from the traffic on I-70, and rural highway widening can literally eat up small towns. Residents might reasonably resent the impacts of transportation improvements meant to ease the flow of visitors to nearby resorts.

Highway projects meant to link cities, resorts, and wildlands in the West may very well evoke regional collaborations of small towns in the same way early expressway projects galvanized neighborhood groups, in protest, within American cities.

Resort Sprawl

Like the other geographies described in this book, the resorts boomed in the 1990s as incomes rose and baby boomers, in particular, started not only visiting them more often, but also buying resort real estate.

The "wealth effect" of the 1990s economic and stock market boom had people who were backpacking in the Wind River Range during the 1970s now buying condos at ski areas or houses on golf courses, spending time in developed resort areas instead of in the backcountry. This resort real estate boom did not slow with the economy in the early 2000s; all-time real estate sales records were set in 2005 in classic resorts such as Vail and Aspen.[24] The second-home boom is a national phenomenon, likely to grow even larger as baby boomers move into the prime resort real estate buying age, pegged at 55–64 years by a recent study.[25]

Dramatic increases in investment, real estate appreciation, and demand for services have induced an affordable housing crisis in most resorts (fig. 6.3). Rising prices have forced many workers to live farther and farther from the resort town proper, creating an enlarging sphere of resort influence, now taking in entire subregions of growing bedroom towns. Inevitably, the resorts themselves have also sprawled, not only because workers have moved farther out for cheaper housing,

Figure 6.3 High-end residences overlook worker housing in the valley bottom near Vail, Colorado. *(William Travis photograph.)*

but also because the resort boom itself could not be contained by the pinched geography common to many resort towns. Klingenstein observed resort cores diverting many developments out to their rural fringes, especially year-round residential and service land uses, but also eventually high-end, second-home developments and even entire ranches purchased as amenity retreats (see chapter 7).[26]

The resorts, therefore, present a three-part geography:

1. The core resort, and adjacent facilities such as ski areas and golf courses
2. The staging areas (and the roads and highways that link them to the resort), in outlying towns and rural areas, to which many resort workers as well as commercial land uses that don't suit resort style (e.g., self-storage facilities; roofing, plumbing, and landscaping companies) are banished
3. A sprawling amenity zone, something like an exurb, including high-end residential developments (often large-lot, large-house second-home developments), outlying golf courses, fishing properties, and, more and more, trophy ranches. In places this zone envelops the towns of the staging area, thus raising their real estate prices, too.

The "Down-Valley" Shuffle

It now seems inevitable that wage earners would be squeezed out of booming resorts, but as places like Park City, Crested Butte, and even Aspen first began to boom, it appeared that the "locals" would benefit through job growth (which increasingly included high-paying professional jobs and demand for skilled craftspeople), home appreciation, and investment opportunities. And many did. But the growing cadre of workers, most of whom arrive too late to invest in, or will never have the money to buy into, the resort phenomenon, as well as the longtime residents who find it increasingly difficult to live a normal life in a resort setting, eventually head "down-valley," often, as historian Hal Rothman and journalist Raye Ringholz documented, quite disenchanted with what their towns have become.[27] The phrase "down-valley" is derived, of course, from the geography of the West's ski resorts, which are naturally at higher elevations, at the tops of valleys, but it applies to the exodus of residents in any direction.

Towns down below may provide cheaper housing (fig. 6.4), such as Basalt–El Jebel–Carbondale below Aspen; Bellevue and Carey below Sun Valley; and even Carson City, Nevada, down the east side, or Colfax, California, down the west side, of the Sierra Nevada from the Lake Tahoe resorts. But "down-valley" can also mean uphill; workers in Vail (at roughly 8,000 feet elevation) commute to Leadville (at over 10,000 feet) for housing. It can also mean over the hill: some Breckenridge, Colorado, workers commute over Hoosier Pass to Alma and Fairplay, and workers in Jackson commute over (the very steep, often snowy, scary, avalanche-prone) Teton Pass into Victor and Driggs, Idaho (fig. 6.5). More recently, Victor and Driggs have become resortified themselves, and workers have been pushed even farther out. The lifeline of U.S. Highway 89, blocked off and on by landslides in the Snake River Canyon, now connects them to Alpine, Wyoming, the new emerging Jackson Hole bedroom community. The same dispersal happens everywhere resort real estate appreciates: workers spread out from the ritzier desert golf communities around Palm Springs; out of Sedona into the small towns in the Verde Valley; and east from Park City to Oakley, Woodland, Kamas, and Coalville. Even many western towns that are not classic resorts, but are still gaining cachet and becoming second home–worthy, such as Bend, Oregon, and Prescott, Arizona, are witnessing worker flight to cheaper digs.

But resort sprawl also means that cheaper housing is not guaranteed, even far from the resort core. Business journalist Jason Blevins captured an important resort geography trend in his October 1, 2000, *Denver Post* article "Downvalley Goes Uptown":

> Real estate prices are soaring in tiny communities located downvalley from nearly every major ski resort in the nation. Quiet hamlets that once provided inexpensive living for resort workers and ranchers are becoming the playgrounds of wealthy investors.

He cited high-end developments such as River Valley Ranch in Carbondale, Colorado, and Teton Springs Golf and Casting Club near Victor, Idaho, that signaled the arrival of resort real estate in towns that once seemed far removed from the glitz of Aspen and Jackson. In business journalism style, Blevins struck a positive note by saying that this spread of resort development was "infusing a new life into the quiet

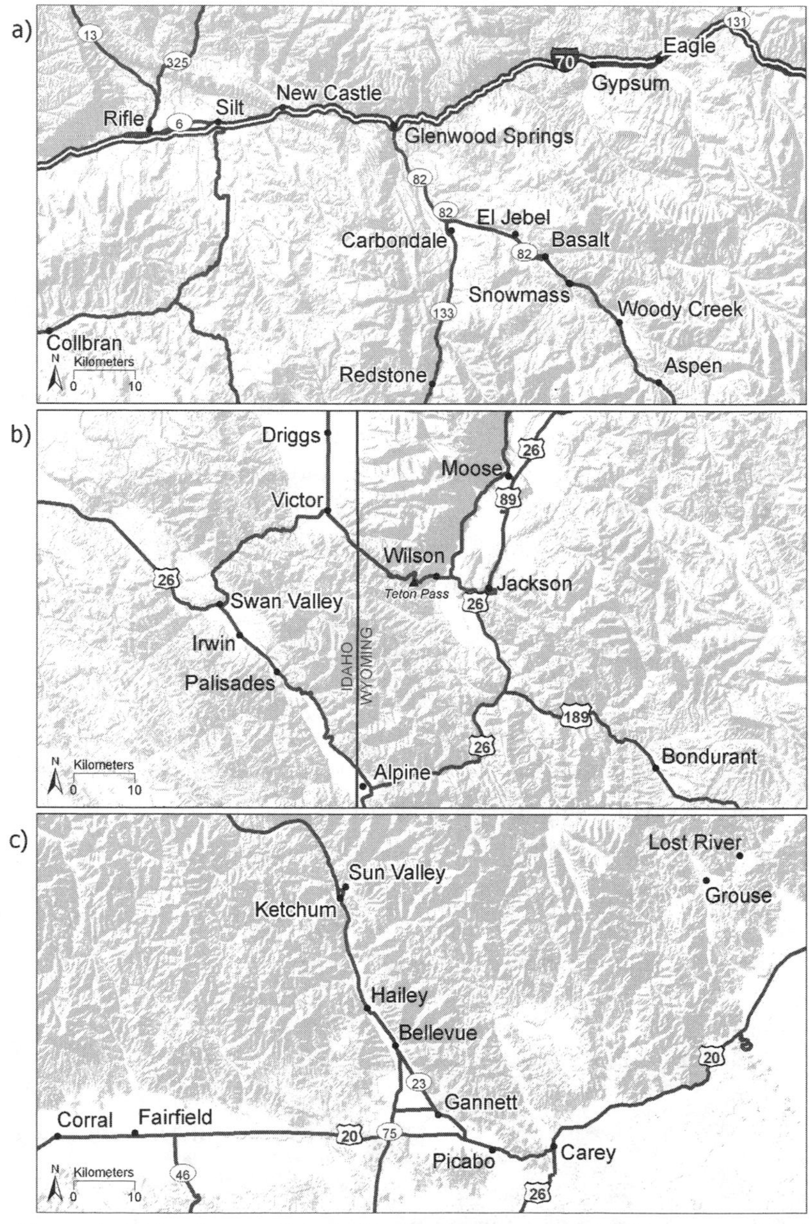

Figure 6.4 The down-valley shuffle: Real estate prices and the geography of mountain resorts force workers far down the valley, or even across mountain passes to distant communities, for housing. Here we map three of these commuter zones: (a) from Aspen to Silt and Rifle (nearby Gypsum is a "down valley" town of Vail); (b) from Jackson to Victor and Alpine; and (c) from Sun Valley to Bellevue and Carey.

Figure 6.5 Avalanche-prone Teton Pass, between Jackson, Wyoming, and Victor, Idaho. *(Thomas Dickinson photograph.)*

hamlets," but he also recognized that the trend bodes poorly for affordable housing: where will workers go?

Grappling with Aspenization

As in so many of the growing cities and resorts in the West, 1993 was a seminal year for Park City. The Interior West's population had grown only slowly during the 1980s, with many rural areas such as Summit County, Utah, actually losing population as the energy boom subsided. Now, however, with an improving national economy and a slumping California economy, the Interior West was suddenly hot property. Resort towns such as Park City, which had actually languished while developers invested in places with more cachet, such as Aspen and Sun Valley, were surprised by their sudden success. Summit County moved to the top of the nation's growth list, vying with Douglas County, Colorado, as the fastest-growing in the country. With some 400 local real estate brokers and developers promoting and relishing the boom, the stage was set for Ringholz's "paradise paved."

The practice of planning had also changed since the early 1980s, and planners like Myles Rademan approached the new times with less planning theory boilerplate. Rademan organized thirty-four "living room" meetings, taking a cue from the focus group approach pioneered by the marketing industry. Next, a citizens' advisory committee was organized as a sounding board. A revised master plan emerged, and went through the standard review and adoption process ("standard" meaning controversial).

The new master plan pointed out an irony, in understated terms: town leaders in the 1980s had sought growth and development and structured the land management code to encourage residential and commercial projects. These pro-development regulations had been, according to the 1993 plan, "largely successful."[28]

They had been all too successful for many residents, and in those living room meetings, Myles Rademan got an earful of complaints about crowding, construction, noise, "hurry hurry attitudes," and the influx of wealth, made obvious by the new monster homes looming on the ridgetops; locals were worried about "Aspenization." Rademan wrote, after the meetings, that they were marked by a surprising "depth of grieving" over longtime residents' lost sense of community and place. Residents lamented the loss of "specialness, separateness," and even the "funkiness" and "grooviness" of the old mining-turned-ski town. But Rademan sees a pervasive "urban denial" in their comments. They want, even still expect, Park City and Summit County to remain rural, despite all they see happening around them, and despite their own demand for urban-style services and amenities. They want to see cows grazing in the pastures around Park City, but appreciate the rise in the value of their own real estate. They don't realize that the cows they see are already an anachronism, grazing on pastures worth at most $500 per acre for agricultural use, but selling by the mid-1990s at $80,000 per acre for residential development. These ironies—what Rademan calls their "cognitive dissonance"—carried over to the residents' proposed solutions. They wanted growth limited, but they did not want property rights infringed upon. They wanted development regulated, but they did not want local government to tamper too much with the free market. And they wanted a healthy and diverse community, but they didn't want to subsidize worker housing.

Glimpsing the Resort Future

If Aspenization is the future of many, maybe most resorts, then what about the future of Aspen? It is certainly a mature resort in any conception of the resort development cycle, but in many ways, not always obvious to the distant observer, it is constantly changing, and may still be a bellwether for other resorts.

Three questions must be answered to chart Aspen's future, as well as the evolution of all western resorts. First, how will recreational and resort tastes, especially among the affluent, change in the future? (Could they change away from the Aspen type of resort?) We already know that smaller proportions of young people ski, but might they still be interested in a vacation home in a pleasant and exclusive mountain or seaside town?

Second, like many economic, real estate, and recreational patterns in the United States, the future of Aspen and other resorts depends somewhat on demography, especially future retirement patterns. Resort planning consultants like to show the population pyramid of the United States, arguing that the baby boomers are just getting into their highest-earning, and highest-investment, years. But as the baby boomers reach retirement (in large numbers starting just after 2010), will they retire to places like Vail or Palm Springs? What is the likelihood that Aspen's "second homes" will become "first homes" at retirement? A recent study in Vail suggests that many second-home owners plan to retire there permanently. I am skeptical, although maybe emerging patterns of earlier and younger retirement will create a retirement boom at least through 2020 in the resorts. After those retirees get older, perhaps we will see disinvestment in resorts that are too cold, too hot, or too far away from good medical facilities.

Third, how will resorts collaborate with nearby communities and adjacent jurisdictions, onto some of which inevitably fall the burden of housing and providing vital social services to resort workers? Many resort town officials and business leaders already collaborate with other resorts, sharing stories and experiences, what worked and didn't work. But as useful as it is for Aspen to share experiences with Jackson, it is more important for Aspen to work with Silt and Rifle, the towns 60–70 miles away to which some of Aspen's workers head home every night, and for Jackson to work with Victor, Driggs, and Alpine, for the same reason. Here's a place for regional planning and for "councils of governments" like the Northwest Colorado Council of Governments, arrayed around resort areas but also including the resort staging towns.

Another great challenge facing resorts is their ongoing, worsening housing affordability crisis. Can Aspen continue to supply its residential property owners with the high-end personal services that make it a desirable resort community? The conundrum for Aspen, Pitkin County, and the entire Roaring Fork Valley is how to maintain its "attractive exclusiveness" while sustaining the workforce needed to fill the increasing demand for personal services in both the commercial and residential sectors.

Getting Aspenization Right

There is some logic to using Aspen's place name as a verb: it set the pattern for the second-home boom, the astronomical rise of resort real estate prices, and the challenges of catering to an increasingly

Figure 6.6 Affordable housing near Aspen, Colorado. *(William Travis photograph.)*

wealthy clientele. But Aspen and Pitkin County have also led the West in terms of planning solutions, including innovations in affordable housing and transit (in Aspen and up and down the entire Roaring Fork Valley); impact fees; transfer of development rights; commercial development code; restrictions on house size; and even parking. Aspen, more than any other resort, has tried to solve the affordability problem, using various policy tools aimed at housing at least 60 percent of its workers in town (fig. 6.6).[29] The City of Aspen and the Aspen Ski Company have recently joined forces to address global warming, which obviously threatens the basis for all ski towns, and set a new standard for assessing a community's carbon footprint by including the carbon dioxide emitted by visitors traveling to Aspen.[30]

Aspen folded long-term sustainability into its latest community plan, creating an Economic Sustainability Committee in 2002, which identified several important trends that would affect Aspen businesses in the foreseeable future (which I paraphrase here):[31]

- The maturation of the ski industry (and its participants): Ski resorts have invested significantly in improved facilities to keep a flat base of skiers, often paying for the investment with their real estate profits. Aspen believes that it has not stayed competitive in this arena and needs to grapple with the new realities in the industry.
- A shift from a ski-driven to a real estate–driven business community: Real estate now dominates the economy, with the second-home market dominating the real estate sector, along with a more recent transformation of short-term lodging into interval ownership (time-shares).
- A saturation of the retail sector by establishments selling expensive goods, causing retail space rentals to surge so that only the very high-end companies can afford to operate in town.
- Increasingly vigorous nonprofit, cultural, and think-tank institutions: Aspen has long been a locus for nonprofit social institutions such as the Aspen Institute, Windstar, and the Aspen Global Change Institute. Their activities and budgets have grown, and they have become part of Aspen's attraction.
- Increasing demand for social services, such as mass transit, employee housing, and child care.

Aspen's plan for responding to these trends is, broadly speaking, to make the community sustainable for permanent residents, second-home owners, and visitors alike—that is, to achieve Klingenstein's hypothetical ninth phase of resort development. One idea is to provide more "affordable retail space" that could not only offer the conventional retail services that residents need, but could also maintain an opportunity for locals to start and run businesses. Add zoning changes that ease downtown redevelopment, support the growth of lodging, and encourage affordable housing, and Aspen is making progress toward a more diverse, sustainable resort community.[32] The city and county cooperate on land use, including an urban growth boundary and transfer of development rights program, and both work with the regional communities on transportation, water resources, and social services. In a way not just Aspen, but the entire Roaring Fork Valley, offers a glimpse of the net phase in resort geography.

Other resorts have tried to emulate Aspen's success as a think tank and crucible for global policy development. Jackson Hole has visions of becoming the "Geneva of the Rockies," a meeting place for political leaders from around the world. But the leaders of the Jackson Hole Chamber of Commerce have trouble getting local merchants on board: will Geneva pay their rent in an ever-appreciating real estate market?[33] They will need to emulate Aspen's other successes: building a community among businesses and people and working within a regional setting. Maybe to "Aspenize" is not all that bad an idea.

Plate 1 Western development is often constrained to valley-bottom locations by public lands, transportation infrastructure, and the attraction of riparian settings. This view shows the finger of development extending from the Sun Valley ski resort (*lower right*) down to the Snake River Plain (*upper left*). *(USDA Farm Service Aerial Photography Field Office photograph layered on U.S. Geological Survey land use and land cover data.)*

Plate 2 The suburban edge is often marked by an interlacing of developed and less transformed landscapes. *(John Fielder photograph, used with permission of the Colorado Sprawl Action Center.)*

Plate 3 Golf course greens stand out in a dry landscape. Recreational development is also transforming the West. In this case, an irrigated ranch has been transformed into an irrigated golf course. *(William Travis photograph.)*

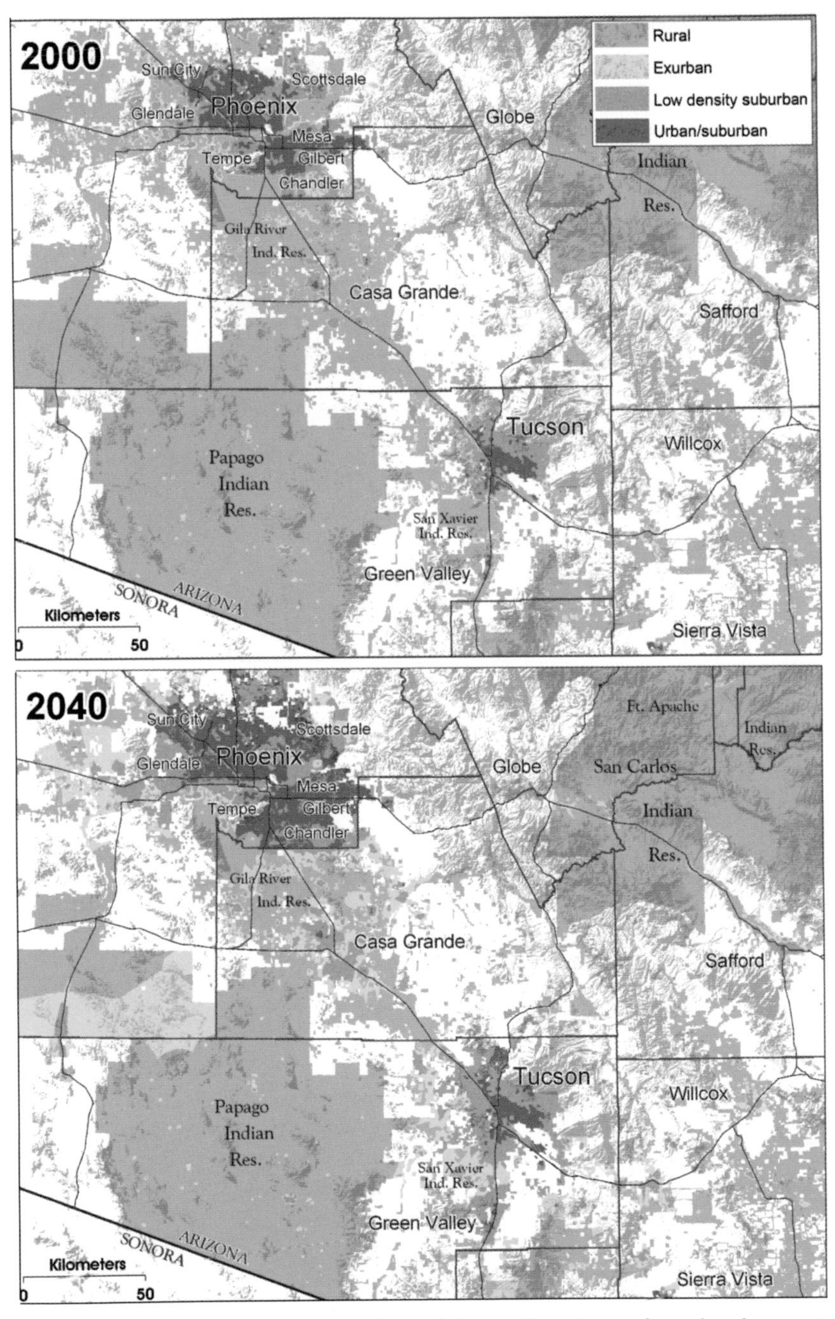

Plate 4 Southeastern Arizona's footprint of urban-to-exurban development, 2000 (actual) and 2040 (projected). Recent news articles about the convergence of outlying Phoenix and Tucson suggest that this projection may be conservative.

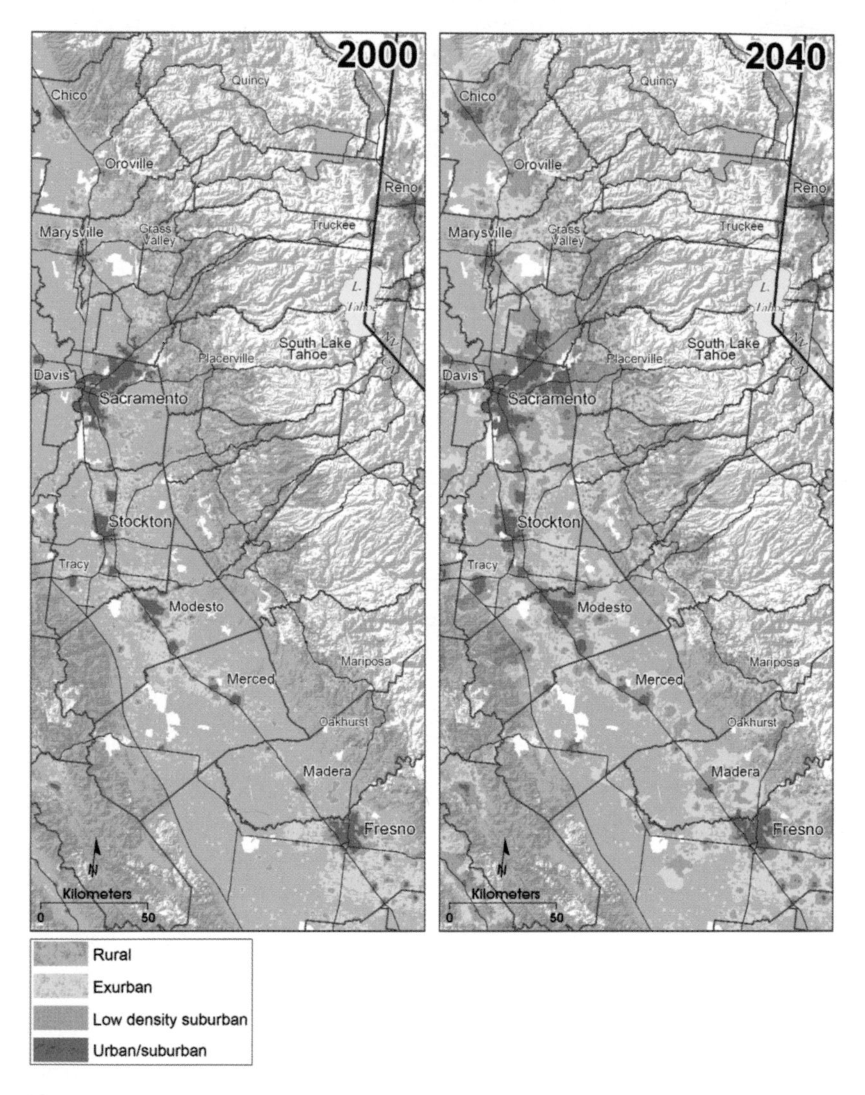

Plate 5 The extensive exurbanizing zone in the Sierra Nevada foothills, 2000 (actual) and 2040 (projected). Fingers of exurban development become large swatches over time, especially in the foothills north and east of Sacramento and up the Merced, Toulomne, and Stanislaus drainages above Modesto.

Plate 6 The typical mountain resort town insinuates dense development into forested and near-wilderness settings. *(William Travis photograph.)*

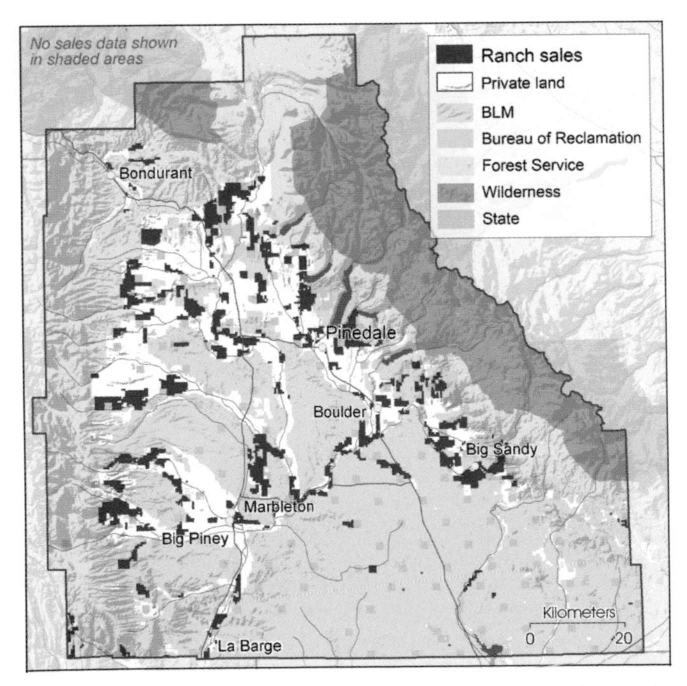

Plate 7 Footprints of ranches sold between 1990 and 2001 in Sublette County, Wyoming. Ranches along streams were especially marketable, as were those along the border of national forest lands on both the east and west edges of this part of the Upper Green River valley.

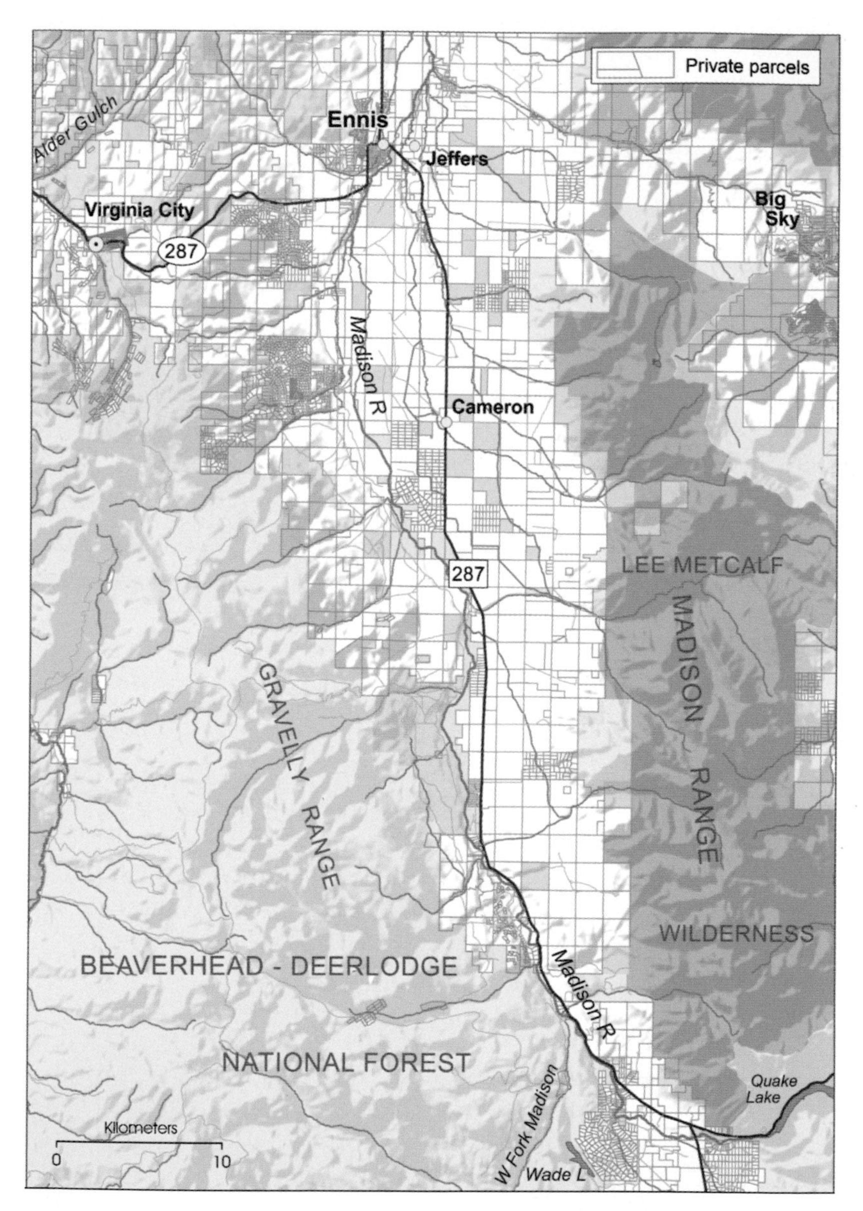

Plate 8 Land parcels in Montana's Madison Valley illustrate the variety of land ownership patterns that evolve over time on the gentrifying range: large ranches; ranchettes, and large residential lots.

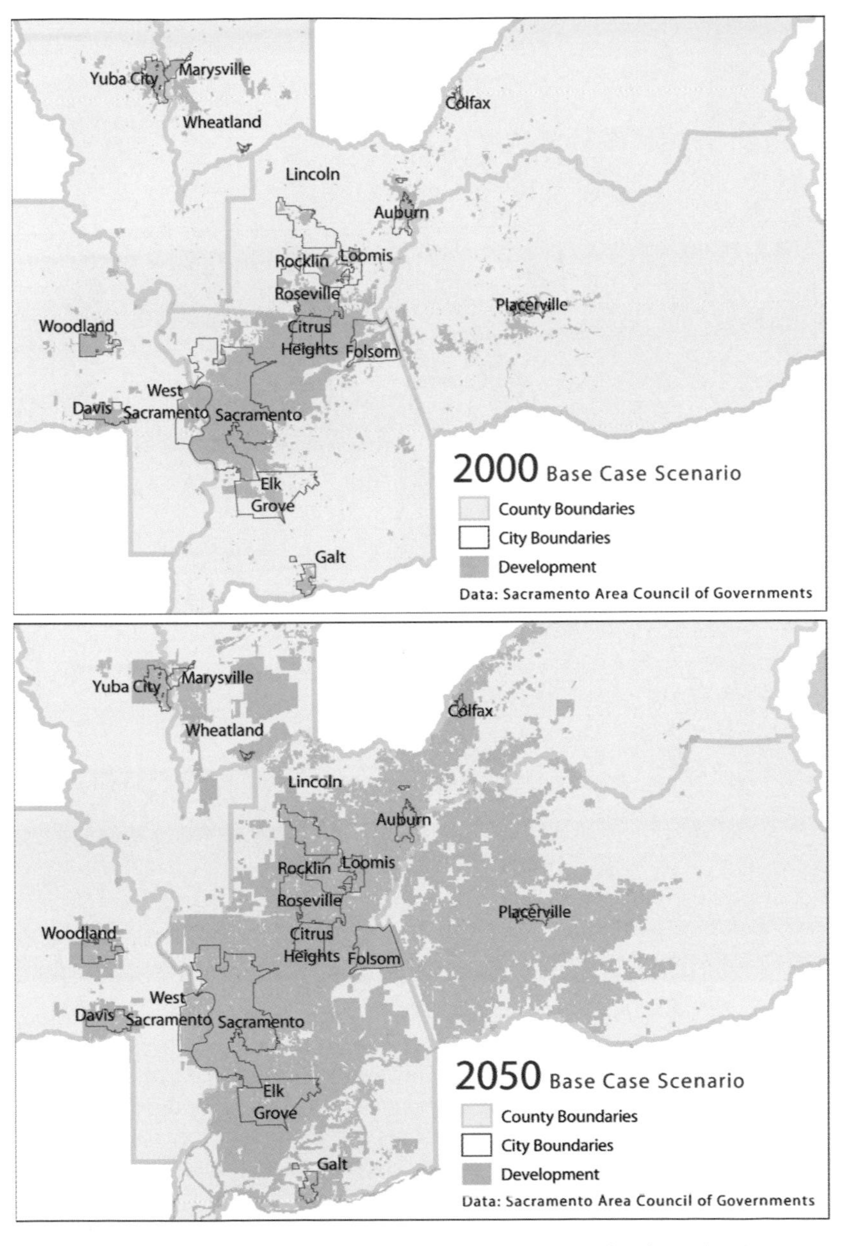

Plate 9 The current and projected future footprint of urban development around Sacramento, California. Scenarios like these allow residents to better envision alternative futures and their cumulative effects. These projections were made with PLACE³S planning software. *(Sacramento Council of Governments, used with permission.)*

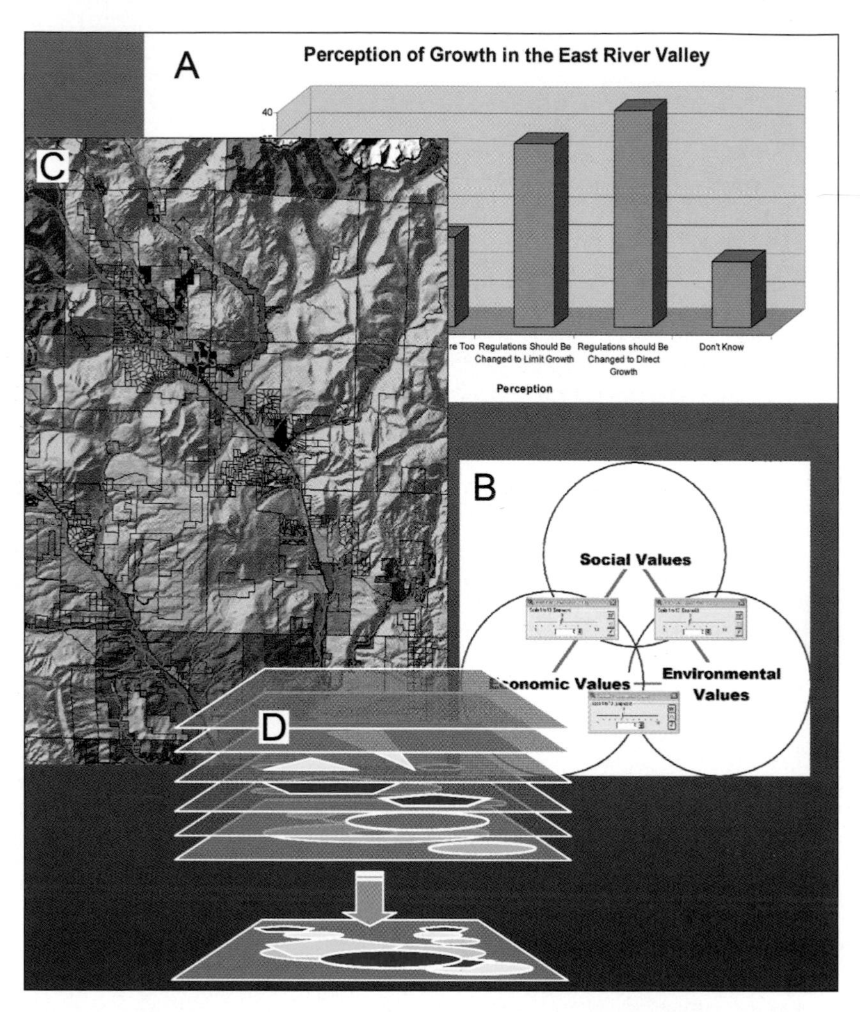

Plate 10 CommunityViz planning software applied to the Gunnison, Colorado, comprehensive plan update. The model allows participants in the planning process to (a) collect community attitudes about land use and planning needs; (b) establish social, economic, and environmental values for land; (c) map those values onto current land ownership and use; and (d) identify areas and planning approaches that could yield the most desirable land use outcomes. *(Placeways, LLC, and the Orton Family Foundation.)*

7 The Gentrified Range

New Owners of the Purple Sage

BEYOND THE CITY CORES, their surrounding suburbs, and the far-reaching exurbs lies a western rural geography that provides the broad matrix for the other development landscapes and creates a development landscape of its own.[1] These rural open spaces, mostly treeless, constitute the great western "range." Out of this landscape rode John Wayne and other cowboy heroes of film and television. Every American (and many others around the world) even minimally exposed to "Westerns" carries a mental image of rangelands as the archetypal western landscape: sparse vegetation backdropped by a panorama of buttes and mountains. From a land use perspective, the range was to be settled and put to work; its settlers aimed to wrestle something wild and inchoate into something that produced value.[2]

The combination of history, economy, and landscape in the rural West creates a unique sense of place. The mythic grandeur of the western range adds enchantment to the utilitarian incarnation of rangelands for cattle production, as does the image of the cowboy and the rancher. Ranches, the largest private land parcels in the United States, were carved out of the public rangelands during the homesteading era and took their place in the commodities chain along with the timber mills, mines, and oil patches. But the "ranch" took on more complex, antiurban connotations, meanings that became especially attractive to outsiders as the West's rural economy was restructured and extractive

Figure 7.1 New homes spring up on the range as ranches sell. *(Julia Haggerty photograph.)*

industries were eclipsed by the consumption of landscape amenities.[3] These ranches, originally sites of commodity production, became a commodity in their own right, purchased for recreation and for the "ranch" ideal.

Non-ranchers are buying the range (fig. 7.1). Members of a new landed and mobile gentry, attracted to the imagery of the Old West, the region's fetching landscapes, and the cachet of owning a piece of it, are buying new homes on the range, either subdivided ranchette properties (well away from town and used as a recreational parcel or second home) or entire ranches. The new land uses they bring to the western range have some trappings of pastoralism—horses and cows may still be involved—but nonagrarian economic forces and preferences dominate.

The pattern and process of this transformation of the range will determine the look of the rural West and the future of its rural communities. Seasoned range scientist Jerry Holechek took a tour of the Interior West in the summer of 2000 and found "western ranching at a crossroads."[4] Increased demand for rangeland for alternative uses was first on his list of "challenges" to western ranching:

> Improved communications, rising affluence, and demographic shift have created tremendous demand for western rangeland as home sites and ranchettes. This has elevated private rangeland property values several fold over what they would be merely for grazing. Rangeland fragmentation into home sites and ranchettes is now ubiquitous in several western states. In areas such as central Colorado, central Texas, southern Arizona, central Utah, and central Oregon former ranching economies have been almost completely transformed into seemingly endless tracts of low density urban sprawl. (17)

As Holechek observed on his sojourn through the West, one result of the gentrification of ranchland—and the sale of ranches to nonranchers—is subdivision, not usually into suburban-sized lots, but into 40–160-acre ranchettes and hobby ranches. Most buyers cannot afford a thousand-acre spread, and probably would not want to manage that much land. No one has yet conducted a region-wide assessment (Holechek called for one, citing the need for better quantification of land use change), but anecdotal evidence and case studies suggest that large-lot ranchland subdivision is spreading, especially in charismatic landscapes throughout the Rockies, in California's Sierra Nevada, and in the picturesque deserts of Arizona and New Mexico.[5]

Western rangelands also now attract nonagricultural buyers who want the whole ranch, or maybe even a couple of ranches. Although ranches have always been of some interest to wealthy folks looking to own a piece of the West (the original "dudes"), the pace quickened in the 1980s and 1990s with purchases by several high-profile buyers, including Ted Turner, David Letterman, Tom Brokaw, and a host of movie stars. In its 1995 special issue on "The Rich," the *New York Times Magazine* proclaimed that "a ranch in the Rockies is now one of the proudest trophies of the rich."[6] This land avoids subdivision because the new owners buy land precisely to own a ranch: not for production, not for development, but for recreation, privacy, bragging rights, and maybe for conservation. They buy the ranch as a hobby or a vacation spot; they are "amenity ranchers."

Both of these trends are part of what some developers are calling the "New Ruralism," a slightly tongue-in-cheek reference to New Urbanism, except that New Ruralism spreads development out rather than concentrating it. The trend is national: the nation's largest rural land developer, the St. Joe Company, is marketing lots of 5–150 acres in Florida's pine forests, far from any city, to "people who have always wanted to live on a farm but don't see themselves as farmers," people who would use their rural homesteads for "dabbling in horse riding, beekeeping, wildflower growing and field plowing."[7]

The End of the Rural?

For obvious reasons, scholars link rural landscapes with agrarian production and culture. Much of their understanding of rural geography and economics focuses on the farm, agricultural markets, and farm labor. In particular, they have tracked the decline of the family farming structure that once dominated rural land use in the United States, a decline they believe to be associated with the ascension of the "agribusiness complex" and the low returns on labor and investment that accrue to the primary producers: the ranchers and farmers. They link almost all the other problems that some rural areas face—lack of commercial and retail services, unemployment, youth out-migration, unrelenting poverty, and small-town decline—to the persistently depressed agricultural economy and the never-ending "farm crisis" (or to declines in other rural resource sectors such as mining and timber).[8] Thus the study of the rural has traditionally been the study of farming and ranching, even as urban professionals bought farms, remodeled or scrapped the farmhouses, and commuted to the city (or brought jobs and income with them, along with urban values, *Green Acres* style).

In 1980, urban geographer Pierce Lewis argued that there was little or no nonmetropolitan geography left in the United States: "the metropolis is almost everywhere," he said. Dispersed transportation and communication systems had devoured most of the cultural and economic distinctions between, say, New York City and Miles City, Montana (his examples!).[9]

Lewis was a bit ahead of his time, but his assessment of America's rural future was on the right track. The distinction between urban and rural life has declined dramatically, except in areas that are mired in the permanent agrarian slump, such as parts of the Great Plains, the

South, and Appalachia.[10] Technologies barely discernible in 1980 have freed many people from locational constraints. In the 1980s and 1990s, analysts began recording an apparent rural restructuring in which, as geographer Peter Nelson of the New West School observed, "historic ties to the land both through work and recreation are transformed."[11] Rural areas now attract nonrural residents with little or no link to agriculture or to other traditional rural economies. These "neo-homesteaders" choose to live in often very rural settings, acquiring land not for production of material goods, but for lifestyle.[12] Rurality, especially the cowboy, ranching version, has become chic; thus geography and the leisure class have come together to create a new pattern of land ownership and use in the West.

The Gentrified Range

If we assign the term "exurbs" to the far-flung commutersheds of western cities, as described in chapter 5, then another term is needed for the insinuation of nonagricultural residential and commercial land use patterns into deeply rural areas. Although the St. Joe Company and other developers may market the "New Ruralism," I prefer the term "rural gentrification," which captures the key process: the appropriation of rural land with capital not associated with, or earned from, traditional rural land uses such as farming, ranching, logging, and mining. Many Americans with the means are deploying their wealth to realize a dream of owning a piece of the West, and they are buying these lands precisely for the amenity of rurality. Some buy 40 acres; others buy 4,000 acres.

"Gentleman" or "hobby" farmers and ranchers, who by definition have another, nonagrarian career, have owned ranches in the West for a long time. By the 1880s, a wealthy class, including dudes from Europe and the East (such as Teddy Roosevelt), interested in horses and hunting retreats, were buying ranchland in the West.[13] But these early amenity ranchers were scarce; they tended to cluster in certain areas, and as geographer Paul Starrs has shown, they were also investors who actually helped create large working ranches.[14] Several characteristics of today's rural property market mark the new homesteaders as a different strain of owner. Hobby operations of the past clustered around sizeable towns that provided a career and income base.[15] The rural gentrification that is spreading across the West today reaches into the heart of rural rangelands.

The pattern and magnitude of this land use transition is little known. Data on rural land sales and subdivision are kept at the county level, and few studies have taken a broad view. A 2002 study suggested that roughly half of the West's ranches are now "hobby" operations, owned mainly for their landscape amenity, lifestyle, and investment value rather than for livestock production.[16] But the geographical outlines of the gentrified range remain poorly mapped.

The Driving Forces behind Rural Gentrification

The forces that drive full-time or part-time residence in rural areas in the mountain and desert West, and that shape the resulting landscape, are at least conceptually well understood. Several studies show that a mixture of economic and quality of life considerations attracts people to amenity-rich rural areas, and that this attraction has been especially strong in the West for the last two decades.[17] Two studies, over two decades, of in-migrants to Montana's Gallatin Valley, where Bozeman is the anchor town, concluded that natural amenities and recreation opportunities, and not necessarily job prospects, attracted both rich and middle-class residents.[18] But what impels newcomers to take on the ownership and management of a significant parcel of land, or even a full-fledged ranch far from town?

Although many studies have set out to understand why ranchers either sell or stay on their land, few have undertaken to grasp why people buy large spreads of ranchlands for uses other than agricultural production. Perhaps questions of motivation are best addressed qualitatively. Some western scholars speculate that this generation of rural land buyers was influenced by the western popular culture that pervaded their youth. Since early in this nation's history, a hearty western mythology has idealized the pastoral dream of ranch life, perhaps spurring the purchase of rural land. As historian Patricia Nelson Limerick writes, "Cowboys were at their peak of popularity in the movies and on TV in the 1950s and early 1960s. . . . As children, a significant percentage of the baby boomers imprinted on the heroes of the open range."[19] Many of these baby boomers are now living out their dreams as the owners of ranchette-sized "horse properties" or even full-scale ranches. One new owner of a ranch, interviewed by geographer Jessica Lage in California's Sierra Valley, encapsulates this theory: ever since he was a kid, he said, his dream was "to be a cowboy." "I always thought I was born a hundred years too late . . . I lived in a small town in California . . . watching TV and playing cowboys and Indians with all my buddies, and the older I got, the more I wanted to live out in the country and have horses and all that."[20] It also makes sense to hypothesize that owning a ranchette or a ranch is simply the next logical extension of exurbanization: given the ability to do so, some people prefer to live (at least part-time) among open spaces. Finally, there is the explicit goal of incorporating residence and recreation: anyone perusing a recent copy of *Rocky Mountain Farm and Ranch* magazine will see that the ads stress fishing, riding, hunting, hiking, and other leisure pursuits on land that until recently was devoted exclusively to commodity production, which required work.

Certainly the disposition of rural lands into ranchettes or large amenity ranches hinges on larger geographic amenities. The demand for ranch properties is especially strong around preserves such as Yellowstone National Park.[21] Indeed, some buyers are so interested in large, intact properties in these areas that they purchase and combine two or more ranches to create even larger private spreads. But in other geographic locations, as within a couple of hours' drive from Tucson, Arizona, the subdivision of ranchlands dominates.

What are the driving forces behind ranch sales? The standard model explaining why traditional ranches get sold or subdivided (advocated in range science and pro-ranching literature) emphasizes the cumulative effects of three stressors: (1) poor markets; (2) federal crackdowns to enforce grazing regulations and environmental protections; and (3) drought.[22] These rationales ignore a set of political and other economic motivations, as well as familial and cultural circumstances, that encourage ranchland sales and subdivision:

- *Price*: The demand for ranchland by nontraditional owners is strong, and bids routinely come at a price far above its agricultural value. The market is expressing a preference, and holding on to land for its production value when the market has valued it for a higher use is simply inefficient (although many ranchers do just that, at least until another pressure comes into play).

- *Family dynamics*: Analysts who have looked deep into the heart of family ranching find the point of inheritance to be especially vulnerable to sale.[23] Because the parents want to pass on financial resources to multiple siblings, some of whom inevitably do not wish to ranch, the ranch itself must be liquidated. One case, occasioned by a "bitter divorce," was described in detail in the *New York Times*. The Dugout Ranch, on the edge of Utah's Canyonlands National Park, went up for sale when the owners split up and needed cash, rather than land, to divide their property and do right by their two sons. Fortunately for them, as the *Times* reporter described, the ranch "had soared in value as big-money buyers swarmed into the area looking to turn places like Dugout into rustic private playgrounds or patches of sprawling condominiums."[24] Both Christie Brinkley and Ralph Lauren had made inquiries, but The Nature Conservancy, via more than a thousand cash contributions, raised $4.6 million and bought the ranch, following "intensive negotiations among the four family members and their lawyers." Similarly, environmental historian Julia Haggerty, who studied the detailed long-term evolution of family ranches in Montana's Paradise Valley, concluded that, although the amenity market has been strong for a couple of decades and certainly encourages ranch sales, rates of turnover have been driven by family dynamics and financial problems, rather than by environmental or regulatory pressures.[25]
- *Cultural trends*: Individuals, social networks, and popular culture influence patterns of land consumption and land use trends in powerful ways, often dictating which landscapes and which land uses are "hot." For example, in the 1990s, three convergent cultural trends determined the fate of rural land in western Montana. The region emerged as a location (actual or scripted) for several popular films (*Legends of the Fall, A River Runs Through It,* and *The Horse Whisperer*). Ted Turner and Jane Fonda's celebrity, and their ranch in the Gallatin Valley, created an aristocratic cachet around ranch ownership and conservation ranching in particular. Finally, social networks led to agglomeration of amenity ranches in partic-

ular areas. For example, in Sweet Grass County, one amenity buyer from Tennessee spurred friends to buy neighboring ranches, all in a friendly competition with one another to own the most acres or the best fly fishing. Reporter Jim Robbins chronicled, for the *New York Times*, the run on ranches in neighboring Carbon County: "The Great Montana Ranch Rush has begun. Movie stars, film makers, airline captains, captains of industry, songwriters, surgeons, advertising executives and writers—anyone with extra cash has come hunting for a big piece of Montana." For Carbon County alone, he cites a long list of celebrities who now own ranches: "Michael Keaton has a place here. The novelist Tom McGuane has one on the West Fork of the Boulder River. So do Dave Grusin, the musician, and Robert D. Haas, chairman of Levi Strauss & Company. Just north of Yellowstone National Park, Dennis Quaid and Meg Ryan have a house. Jeff Bridges is a neighbor. So is Peter Fonda."[26] Some of these ranch buyers try to maintain working cattle operations; others, like Tom Brokaw, according to reporter Robbins, bought for recreation, trout fishing, hunting, and access to wilderness, not to maintain a working ranch.

Footprint: Two Landscape Outcomes of Rural Gentrification

New ruralites' demand for properties yields two main types of gentrified land uses in the rural West. The most obvious is large-lot subdivisions, or "ranchettes," usually carved out of a larger ranch. Some ranchettes host no agricultural activity; others are horse properties or even small-scale hobby farms or ranches with crops and livestock. Less obvious, but equally important to future rural land uses, are large, intact ranches bought by non-ranchers for amenity uses, including the amenity of owning and running a ranch. The first of these patterns fragments land ownership; the second maintains large private parcels, and may even yield larger ownership units. In both cases, some of the new owners take up residence, whereas others remain itinerant, nonresident owners. Together, these patterns affect more land area and more ecosystem types than the cities, suburbs, and exurbs combined, although in a less intense manner. In addition to the footprints of the isolated homes, the transition in ranchland ownership brings large-scale changes in land use as the new owners implement their own ideas about grazing, wildlife, water use, and access. This rural gentrification, referred to as "amenity migration" by some scholars, also changes the politics, economics, and culture of western rural areas in quite profound, but poorly studied, ways.[27]

Subdividing the Ranch

Some rural landowners opt for a large lot with an isolated house: a ranchette. They want some of the qualities of ranch life (isolation, open space, natural surroundings, space to keep and ride horses, and maybe even have some cows) without the responsibility of owning and managing hundreds or thousands of acres. Since the 1970s, and particularly over the past two decades, the average size of agricultural parcels in the Interior West has decreased (a trend that reverses a century of ranchland agglomeration) as more ranchettes have emerged.[28] Large-lot rural subdivision is not an entirely new phenomenon; it has fragmented the rural landscape for decades.[29] But anecdote and case study suggest that it has been spreading especially fast in the West in the last two decades. Regional observers bemoaned the spread of rural subdivision across the West in the 1990s,[30] and the American Farmland Trust estimated that a tenth of the region's best ranchlands will be subdivided by 2020.[31] Geographer Paul Starrs documented ranch subdivision in deeply rural central Nevada, and Jerry Holechek found subdivision surprisingly dispersed across different western rangelands.[32] Significant portions of whole valleys have been subdivided into ranchettes: the Madison in Montana, the Altar outside Tucson, and many of the major valleys in the Colorado Rockies.[33] What drives this demand for rural land?

The national parks and other charismatic protected landscapes were one focus of this development. The best-studied such area is that around Yellowstone National Park, where rural sprawl and ranchette development is filling the small amount of private land adjacent to the park (federal land composes half or more of most Yellowstone-area counties).[34] Rural development has also quickened near other protected western areas, such as the desert mountain ranges in southern Arizona, where a ring of ranchettes is forming around Arizona's Chiricahua Mountains. As The Nature Conservancy reported, "subdivision of rural landscapes is fragmenting and destroying important valley bottom habitat more rapidly than conservation action can be taken to protect key areas . . . [and] habitat connections to adjacent mountains are being lost as traditional ranches are subdivided."[35]

Ranchettes, derided as "too small to plow and too large to mow," form a repetitive pattern across the West mostly as a result of state land use law. In Colorado and Wyoming, for example, lots over 35 acres are not subject to county regulations, so developers design ranchette estates with parcels slightly larger than 35 acres, thus avoiding rules on roads, sewer lines, and other utilities. In Montana, the cutoff is 160 acres (but it was 30 acres for years, and many 30-acre plats remain to be developed). In addition to driving forces such as economic shifts and quality of life preferences, the political culture produces a landscape of extensive lots dotted with dispersed houses.

Ranchettes may be created as residential lots carved out of ranches one at a time. Ranchers often sell off lots piece by piece in order to keep the ranch or farm in business; over time, these one- or two-lot subdivisions, apparently minor impositions, can fill a ranching landscape with isolated homes. The most extreme form of ranchette development is the collective subdivision of an entire ranch all at once, in which some 500 or more acres may be parceled into lots of 30 to 100 acres, each with an isolated home. One 700-acre base property studied by geographers William Riebsame, David Theobald, and Hannah Gosnell in Colorado's East River Valley was subdivided into nineteen dispersed homesites, each on slightly more than 35 acres. Negative ecological effects accumulate quickly when over a dozen large homesites, along with 15 miles of two-lane graded roads and water, sewer, and utility infrastructure, replace several hundred grazed acres, occupied only by two houses and a few outbuildings and accessed by a dirt road. The adjacent ranch, of 1,000 acres, was subdivided into twenty-nine ranchette parcels.[36]

The pattern can vary even between counties. In California's Sierra Valley, population pressures from Reno, Nevada, recreation pressures from Tahoe, and interest in what are locally known as "leisure ranches" create the potential for extensive subdivision. The valley is split between Sierra and Plumas counties. Few subdivisions have sprung up in the Sierra County portion of the valley, where the Board of Supervisors is in unanimous agreement on protecting the valley floor: the county's general plan recommends a 640-acre minimum lot size. Neighboring Plumas County, on the other hand, encourages development, and several rural subdivisions there function as bedroom communities for Reno, roughly 30 miles away.[37]

Ranchette developments, despite their low-density nature, can affect the landscape significantly by introducing roads, pets, and people into large areas previously disturbed only by livestock grazing. They consume large amounts of land and require more infrastructure (such as roads, sewers, water, and electric facilities) per dwelling unit than the land uses they replace. A growing literature on the "cost of rural services" concludes that residential developments dispersed across rangelands require more investment in local services (e.g., sheriffs, firefighters, roads) than they generate in new tax receipts.[38]

The ecological effects of ranch subdivision have been assessed only recently, most notably in a series of conceptual and empirical analyses by wildlife biologist Richard Knight and his colleagues at Colorado State University.[39] Knight argues that working ranches are better for the ecology than ranchettes.[40] The bottom line, also bolstered by studies of rural subdivision in the Greater Yellowstone Ecosystem by biologist Andrew Hansen and his colleagues at Montana State University,[41] is that increased human presence and disturbance on subdivided ranches appears to reduce biodiversity compared with that on working ranches, especially by pushing out habitat specialists (often rarer species), increasing habitat generalists, increasing invasive exotics, and altering natural disturbance patterns.

Buying the Ranch

Since 1999, my colleagues at the Center of the American West and I have been tracking changes in ownership of western ranches as part of our Ranchland Dynamics project.[42] Anecdotes about rapid and pervasive changes in ownership that threaten the integrity of much of the West's private open lands prompted us to try to document the trend. Westerners were concerned not only about subdivision, as discussed above, but also about the new breed of owner that was buying—and not subdividing—whole ranches. The bulk of earlier ranch sales were *between* ranching families (which meant less land use change), in contrast to today's predominance of sales to nonagricultural owners.[43]

The prototypical new owner in these stories is an urbanite interested in owning a ranch as an amenity. In her interviews with ranch owners in California's Sierra Valley, Jessica Lage found that new owners are mostly urban transplants who have made their money in nonagricultural professions, as professors, software executives, small business owners, and lawyers. Most do not have roots in ranching, and although they come to it with little knowledge, most maintain livestock production on their ranches. Some are interested in making a profit from livestock—even though ranching is not their primary source of income—either by leasing their land to longtime ranchers or by selling their own cows. For most of these new owners, the appeal of the rural lifestyle is ranching itself, and most are also drawn to the open space and scenery, or to the good hunting the valley's deer herd provides.[44]

Unlike their predecessors, the new owners are insulated from the less idyllic aspects of ranch life: debt, crop failure, and stagnant cattle prices, to name only a few. In addition, according to the persistent narrative in rural areas, they lack the interest and skills to manage the land effectively. Lots of rural land in inexperienced hands is leading, the story goes, to an ecological train wreck. This moral tale, as well as the "death of ranching" subdivision narrative, foretells the end of an era as traditional ranching is squeezed off the land.

In addition to residents of rural areas, conservation groups and ranching organizations have been aware of, and concerned about, the trend of increasing ownership turnover since it first appeared on their radar in the 1980s, but they have few data with which to assess the trend. The Center of the American West's Ranchland Dynamics project set out to gather some numbers on ownership turnover and to understand its consequences. We found that there is, indeed, a significant turnover in ownership under way on the region's ranchlands, especially in charismatic landscapes such as the Rockies, the Sierra Nevada, and some desert areas. Our initial study focused on the bellwether Yellowstone region: rates of ranch turnover varied across the area, but overall, a fifth of the ranches changed hands between 1990 and 2001. Some of the more rural places, such as Stillwater County,

Montana, saw only a tenth of the ranches turn over. In more amenity-driven markets, such as Sublette County, Wyoming, over a third of the ranches, and almost half of the private ranchland in the county, changed owners in the eleven-year study period, some more than once (fig. 7.2).[45] With very few exceptions, these sales were to what we classified as amenity interests, buyers looking not for agricultural production but for recreational land, privacy, and the many other amenities of ranch ownership.

Amenity buyers typically pay much more than what ranchers reliant on commodity markets can afford; ranchlands even in less attractive landscapes across the West now go for twice to ten times their agricultural value. In prime areas rich in trout streams, adjacent wilderness, and dramatic mountain and desert terrain, ranchland can fetch tens of thousands of dollars *per acre*, even when it is sold in units

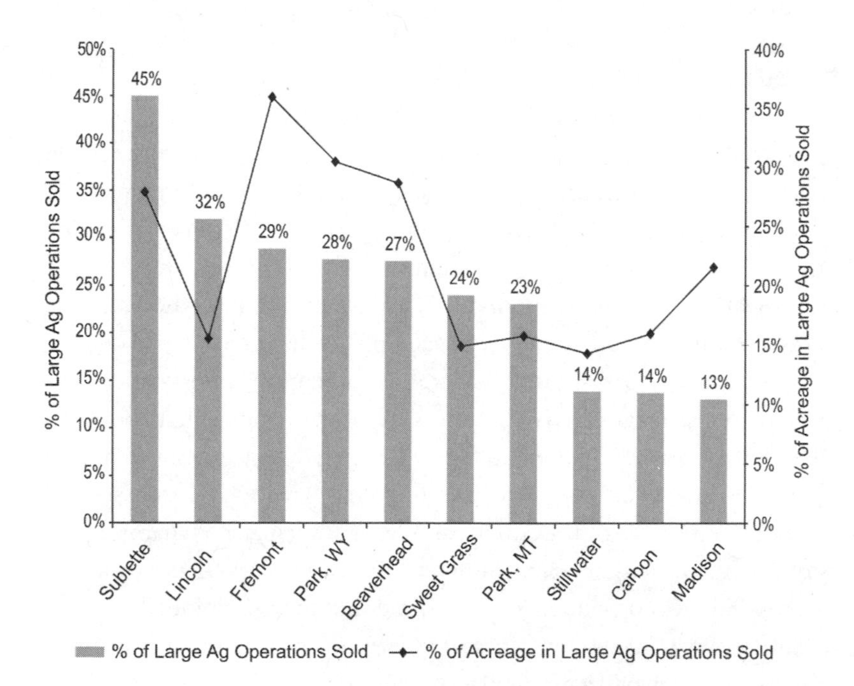

Figure 7.2 Large ranch sales as a percentage of all large ranches and all land in large ranches in the ten Yellowstone counties, 1990–2001. In some areas, like Sublette County near Jackson, Wyoming, almost half of the ranches have changed hands in a decade, most switching to amenity owners. In other areas traditional ranchers hold on and turnover is slower.

of thousands of acres. Many amenity buyers build large, elaborate homes on their ranches, both reflecting and contributing to skyrocketing ranchland prices.

It is difficult to define the geographic footprint of amenity ranching, although a few generalities emerge. First, of course, the land parcels themselves, like all ranches, constitute large blocks of private open land with only limited improvements, typically a few roads, fences, and water sites. The big difference between amenity and traditional ranches shows up at the residence (or residences), which is almost always large and elaborate on amenity ranches. Guest cabins, horse arenas, trout ponds, and maybe even airstrips and hangars also tend to sprout on recreational spreads. Perhaps the more significant footprint of amenity ownership is in less obvious changes in land use: fewer cattle; increased hospitality to wildlife; less predator control; maybe more or less hunting. Geographers Hannah Gosnell and Julia Haggerty and wildlife biologist Patrick Byorth found that new ranch owners in Montana tended to reallocate water away from irrigation to fishponds, in-stream uses, and stream restoration.[46]

Recreational ranching tends to be attracted to areas near national parks, preserves, and other large public land holdings. Amenity buyers also appear to seek out landscapes that are critical to western biodiversity, especially riparian and lower to mid-elevation forest habitats. A map of all ranches sold during the 1990s in Sublette County, Wyoming (plate 7), illustrates two important patterns: sales of ranches that border on the national forest (and thus on the forested mountain slopes) and sales of ranches along the streams and rivers. The strategic habitat importance of these ranches, along with the independence of their amenity buyers from livestock production, holds the potential for a transition to a more ecologically healthy western range. People who buy ranches for the wildlife, scenery, and even seclusion and privacy are likely to institute more ecologically beneficial land uses than would owners trying to produce commodities from the land. Furthermore, buyers able to put together one or more ranches are creating large blocks of uninterrupted open space.

Amenity ranchers are in a financial position to relax grazing intensity, and they are spurred on by their wish to use their land for recreational purposes such as fly-fishing, and often by their interest in conservation efforts such as species protection. Many apply conservation-oriented land uses in place of the land's historical agricultural uses, but few have a comprehensive ecological approach to management—many focus on only a few elements, such as elk or trout habitat. Thus conservation owners per se—with the vision, preference, and resources to turn their ranches into functional nature preserves—are a small subset of the emerging ranch ownership regime in the West. Yet these owners do exist, and some are shining examples of what a new owner can do to reverse environmental decline caused by traditional ranch operations. For example, the new owners of the Sun Ranch in the Madison Valley have put in place scientifically sound programs for native trout restoration, wildlife management (including wolves), and weed control.[47]

Avowed conservation owners can achieve solid ecological benefits, as witnessed on lands owned by The Nature Conservancy (e.g., Red Canyon Ranch near Lander, Wyoming, and the Carpenter Ranch near Steamboat, Colorado, are both well-monitored ecological successes, achieving ecological health along with livestock production), but intentions to create a healthy landscape and habitat do not always result in substantial changes in land management or ecosystem health. New practices, such as switching from cattle to bison or encouraging elk, may not always yield the intended results, for several possible reasons. A common complaint we found among traditional ranchers is that the new owners' "ecological" goals (e.g., "I want to see more elk") lead to new types of land degradation. New owners' naïveté may produce unexpected consequences, such as more weeds, introductions of exotic species, and drying up of wetlands. In a sense, the experiment is under way, especially in those areas where amenity and traditional owners reside as neighbors, but right now we do not have monitoring in place to test the outcome.[48]

Moreover, many new owners do not pursue conservation, simply continuing the previous management practices without considering their effects. They often give decision-making authority to a ranch manager, who may or may not share their interest in and concerns about habitat and wildlife. Some new owners are inattentive to the land. Little time spent at the ranch or disinterest in management practices may cause new owners to neglect the ranch, yielding ecological harm.

Although some conservation benefits certainly accrue as new owners create large ranch reserves and seek explicit conservation goals, questions remain about these owners' long-term plans and their persistence on the western landscape. Cases of new ranch empires built quickly (often disrupting local land and labor supplies) and just as quickly liquidated are not unusual. Nor are stories of new ranchers with outside sources of wealth who become committed to, and well integrated into, local communities. Such community involvement suggests that they are there for the long haul, but their offspring's connection to the property is hard to predict.

A Case Study: Gentrifying Montana's Madison Valley

Montana's Madison Valley is a mixture of large amenity ranches, significant ranchette and large-lot rural subdivision, and a few holdout traditional ranches: a good landscape in which to observe the effects of rangeland gentrification, and the focus of a study by the University of Colorado's Ranchlands Dynamics study[49] (plate 8). The valley illustrates the results of an array of rural development trends: a recreational and retirement subdivision boom that began in the 1970s and continues today; the similarly explosive "river runs through it" boom in properties with trout streams; two cycles of major ski resort development; and a strong demand for trophy ranch properties by the nation's elite over the last couple of decades. The county's plentiful private land inventory along one of the most scenic routes to Yellowstone National Park has made it one of the West's most sought-after addresses by urbanites seeking to own a ranch.

Changes in the ownership of large ranches affected the Madison Valley earlier than most western ranching regions. Indeed, the Madison has shown less ranch turnover during the last decade or so than some neighboring counties (fig 7.2) precisely because most of the large properties are already in amenity ownership. During the 1950s and 1960s, fortunes associated with names such as Schlitz Beer and Perfect Circle Piston Rings underwrote the purchases of properties in the Madison Valley. Ranches such as the Valley Garden Ranch north of Ennis and the Cedar Creek Ranch, Bear Creek Ranch, and Sun Ranch, all south of Ennis, have been in absentee ownership for nearly fifty years. At least two other Madison Valley properties were sold at midcentury to buyers who acquired their ranches with fortunes earned elsewhere, but who made the Madison Valley their permanent residence and ranching their full-time occupation. In addition, the 1970s witnessed the subdivision of entire large ranch properties into ranchettes. Today, the valley primarily features a mix of large amenity ranches, many of them with conservation easements, and several ranchette subdivisions.

The Ranchland Dynamics team found that between January 1, 1990, and December 31, 2001, 38 large ranches (each over 400 deeded acres) were sold in Madison County. The median sale size was approximately 1,997 acres, and the average was 4,820 acres, much higher than in any other county we have studied, due to a couple of sales in the neighborhood of 25,000 acres and a couple in the 15,000-acre range. The Ranchland Dynamics team thus estimated that 13 percent of the current large agricultural operations (294) changed hands in the decade of the study. In terms of land, some 183,159 acres—approximately 22 percent—of the land in large agricultural holdings changed hands over the decade.

We applied a typology to characterize buyers of ranches as "traditional ranchers," "part-time ranchers," "amenity buyers," "developers," "investors," "corporations," and "conservation organizations" (table 7.1). Ranch buyers in Madison County during the 1990s and early 2000s were mostly out-of-state amenity buyers, continuing a trend established as early as the 1950s in a few western locales. Twenty-five of the thirty-eight sales (65 percent) were to amenity buyers. The next most common buyer types were investors and developers, with five (13 percent) and four of the sales (11 percent), respectively. Only one sale was to a traditional rancher. We were unable to type three of the buyers.

Some traditional ranches hold on, but those with any future at all are those that have gotten bigger in the past twenty years. Although 100-head, 640-acre ranches could support a family in the 1940s and 1950s, today their owners have either sold out or remain on their

Table 7.1

A typology of ranch buyers in the west.

From the Center of the American West's Ranch Dynamics Project, University of Colorado (see http://www.centerwest.org/ranchlands)

Traditional rancher Generally a full-time owner-operator raising livestock for profit without the aid of a ranch manager; may engage in some off-ranch work (or on-ranch work unrelated to livestock; e.g., outfitting), but derives the majority (or at least in many years a significant portion) of his or her income from the ranch.

Part-time rancher An owner-operator who often has a full-time job off the ranch; ranch income is generally less than the off-ranch income; usually smaller operations.

Amenity buyer Purchases a ranch for ambience, recreation, and other amenities, not primarily for agricultural production; often an absentee owner; may have some interest in ranching but generally hires a ranch manager who makes most day-to-day decisions and does the majority of the work; or, the owner may lease the majority of his/her land or cattle to a "real rancher." The majority of amenity ranchers' personal income is, by definition, from off-ranch sources; the economic viability of the ranch is usually not an issue.

Investor Buys primarily for investment, often with intent to resell in the short term.

Corporation Typically purchases a ranch to function as one unit in a large network of related operations and holdings elsewhere; the ranch is operated by a manager.

Developer Buys the land with intention to subdivide and sell off to others, with profits from that sale the main objective.

Conservation organization Purchases a ranch with intent to manage for habitat, wildlife, etc.

Other Includes state and federal land management agencies, churches, independent loggers, etc., who have a variety of reasons for buying a ranch.

properties as part-time ranchers, in semiretirement or working full-time off the ranch. These trends, together with a strong demand for recreational ranches, combine to make traditional ranching operations increasingly the minority in the county; only ten or so families actually base their living primarily on production agriculture.

The owners of the 18,000-acre Sun Ranch in the Upper Madison Valley are among the most visible "conservation ranchers" in the Rockies. Since acquiring the ranch in 1998, they have embarked on an ambitious program focused on restoring the health of their ranchlands and on exploring ways in which economically viable ranching practices can enhance and complement ecological processes. The corresponding changes in operations on the ranch have been significant, including altered grazing practices on national forest land, ending all irrigation (and donating water rights to Trout Unlimited), tolerating and adapting to the presence of a wolf pack, removing some fences permanently and replacing others with wildlife-friendly fencing, and imposing a five-year-long rest period on large portions of the ranch's range. In addition to placing a conservation easement on the ranch with The Nature Conservancy, the owners have been instrumental in TNC's Conservation Beef program, aimed at cultivating a national market for beef raised in a sustainable fashion. Their parallel commitment to making the Sun Ranch an open laboratory for progressive, environmentally sustainable ranching techniques renders their ranching strategies more transparent than those of the average ranch, possibly increasing the likelihood of adoption of conservation ranching strategies elsewhere.

Despite the attrition of the valley's traditional ranches, Madison County is home to one of the region's most important local ranching advocacy groups, the Madison Valley Ranchlands Group (MVRG). The original impetus for the organization was the widespread transition, in the 1970s and 1980s, that was quickly turning many of the area's large ranches into islands in a sea of large-lot subdivisions. Its founders were interested in identifying creative ways to encourage the viability of traditional ranch operations, but MVRG also discovered the value of establishing common ground with the absentee, nontraditional owners of some of the area's largest ranches. Many of their current activities exploit the synergy between the two groups of ranch owners.

The MVRG has also focused on land use planning, on stewardship and weed control, and on marketing. The group's director reports that programs specifically geared toward benefiting working ranch operations have been especially challenging. An attempt to establish a kind of private grazing lease clearinghouse in which MVRG links up ranchers in need of grass with landowners in need of pasture management, for example, has faltered, in part because of the difficulty of working with homeowners' associations (in which consensus about land management can be virtually impossible). One homeowners' association has gone so far as to prohibit grazing outright within the boundaries of its large ranchette development.

Madison County is still undergoing active land development. Conservation easements on large recreational ranch properties exist side by side with some of the West's most expansive rural sprawl. Rarely have Madison County subdivision schemes proved to be bad investments for developers. Even in the Raynolds Pass area, for example, where a spate of overambitious subdivision in the late 1970s created a temporary glut in the market during the 1980s, new log homes are springing up on the landscape, built on parcels that became hot properties in the 1990s.

Madison County ranked in the top tier of the counties we studied in terms of ranchland acquisition both by amenity buyers and by developers, a combination of trends—also present in Sublette County, Wyoming, and Park County, Montana—that speaks to a sustained demand for recreational ranch properties of all sizes, yielding parallel evolution of extremes: large ranch conservation and ranchland fragmentation.

The West's New Pastoral Landscape

The gentrification of the western range is not so different from past western migrations, each a rush to own and consume the benefits of a charismatic landscape. A combination of individual preferences and economic changes drives rural subdivision as well as the countervailing market for intact ranches. The long-standing affinity of American culture for the pastoral life (even among urbanites and suburbanites) draws people out of the cities and into rural areas, and the broad-scale shift in the late twentieth century to an economy in which western land is less valuable for productive uses than for amenity uses further propels land conversion, subdivision, and gentrification.

The consequences of ranch ownership turnover are not set in stone. The new rush to own a western ranch does not herald the end of commodity production on all ranches (although it will result in some reduction of total production), nor is it an inevitable conservation train wreck; indeed, many new owners are shifting resources into habitat improvement and preservation. The outlines of the newly gentrified range, whether it remains intact or is subdivided into ranchettes, are still unclear, and its land use patterns are still amendable through efforts by ranching advocates, conservationists, and rural communities.

Ranchers, and many environmentalists, argue that economic and regulatory pressure on ranching leads to the subdivision of ranchlands and loss of important open spaces (the "cows vs. condos" debate). I've engaged in this debate myself with ecologist George Wuerthner, an ardent antigrazing activist who takes the extreme position that even the subdivision of all ranches in the West would be ecologically acceptable if it removed all cows from public rangeland.[50] I disagree, but must admit that we don't have the tools to calculate the full ecological effects of large-lot subdivision, much less to compare them to the benefits that might accrue if grazing were abolished on public lands. Furthermore, the subdivision of all ranches in the West is unlikely in any future scenario that I can conjure up, especially given the strong market for intact ranches, although there are places where it has proceeded quite far. Perhaps the Madison Valley is the future of the gentrified range: a small cohort of traditional ranchers and a new breed of amenity and conservation ranchers working among ranchette estates and all striving for a sense of the Old West in a very New West setting.

Part Three

Shaping the Future Geographies of the American West

The development patterns analyzed here presage trends that will mark much of the twenty-first century. Can we put western development on a trajectory more appropriate to the region's values? Traditional land use planning has done little to mitigate the negative effects of rapid western growth; indeed, planning in the West is mostly about encouraging and enabling growth and land development. Yet concern over growth has itself grown, and more than at any time in the region's history, the toolkit for land use planning is rich with technologies, best practices, and organizations that can change the future. Land itself may be private, but the communities and landscapes in which we live are a commons in which we all have a stake, and the democratic practice of land use planning, promoted in a way that engages citizens directly in decision making, offers a way to change the West's future geographies.

8 Understanding the Challenge of Land Use Planning

WESTERNERS ARE FRUSTRATED with land use practices that make every community look like every other one, that devalue sense of place. They want better development patterns, ones that inspire us, that attract our loyalty and commitment. Can we reshape western development? The answer remains uncertain despite hundreds of local plans, written with extensive citizen input, that express enduring notions of what makes a good place to live. These plans include aspirations for neighborhood character, community separation, transportation choices, commercial services, affordable housing, open space, trails, parks and wildlife habitat, growth rates, and even the community's ultimate size. Such plans face real hurdles, however. Land use policies and planning practices in the West are ill-matched to the forces that yield landscape squalor: leapfrog subdivision and splayed-out exurbs, inefficiently sprawling metro regions, blocked access to trails and open lands, and rural development that interferes with agriculture and compromises natural habitat and ecosystem function.

The voices of citizens demanding smarter growth, although louder in decibels at public hearings, remain weaker than the political and economic power of pro-development forces. This power is captured well in a story, told recently by journalist Jon Gertner, about the national homebuilding industry, which he called the "house-building industrial complex." Leading builders routinely bid tens of millions of dollars for land deals across the country in order to bank thousands of acres for future construction, and they mean to build those projects with little intervention by local communities. As one academic who watches the industry put it, "The large builders have taken the position: we're just going to fight . . . we have lawyers, we have experts, we have money, we're going to buy these tracts of land and fight it out."[1]

The scale of such developments is also a big challenge to traditional land use planning. Planning tools that could encompass the spreading frontiers of development—regional comprehensive plans, urban growth boundaries, and effective statewide planning—have simply not caught on in most of the West (and the limited deployment of such tools—in Oregon and Washington, for instance—is under attack by property rights and antiplanning interests).[2] Something approaching effective regional land use planning exists in only a few pockets of the West.

Private entities might preserve some open space, and innovative developers might create and successfully market a few attractive, dense developments that accommodate growth without sprawling across the landscape. But the bigger picture, the future of the larger West, is rooted in more traditional geographies: in suburban growth, exurban and rural development, and resort sprawl. So land use planning remains our chief means for influencing the evolution of western communities, and it must be strengthened if our places are to retain the qualities that originally attracted us to them. The differences between land use plans and actual outcomes on the ground are too great, revealing to each of us that in the rush to develop the West, we have neglected the institutions and policies that could protect many important community and landscape values.

The Problem with Planning

There is a problem with planning, and we must diagnose it, cure it, and get back to the promise of planning, calling our public institutions back into the service of making western communities good places to live. Land use planning as a public policy endeavor is, in theory, the process through which the values of residents and the goals of landowners and developers are coordinated to achieve desirable community development patterns and to meet citizens' expectations for quality of community life. It is the main expression of government restriction on the unfettered use of private property, and it rests on well-established legal principles. Desired future patterns are determined through a combination of citizen input and the application of best planning practices, and they are expressed in comprehensive plans and associated land use regulations that define how development can proceed. But land use planning is also in tension with private property rights, especially the right to change land from one use to another and to reap the benefits of land appreciation. It is also in conflict with the one-size-fits-all pattern of land development that is remaking the entire American landscape. And because regulation that manages the spread of development is, in effect, growth control, planning is also in a struggle with the abiding American belief that growth is good.

Indeed, planning is a weak force in the universe of factors that propel and shape land development. Its ability to guide how communities develop and how the landscape evolves is weakened by several structural and procedural faults. Planning has, over the decades, developed a split personality. On the one hand, much of what planners do is intended to facilitate, even encourage, land development by enabling market forces and the goals of private investors while promoting the perceived benefits of growth. Much of "planning" is actually local and state government's efforts to encourage and enable growth and development by ensuring that the required infrastructure and services are available and, indeed, that the land is available. On the other hand, citizens look to planning, especially land use planning, for growth management, and they feel ill-treated when the system does its daily job of getting land annexed, areas zoned for development, and development permits approved for projects that often violate what many residents think their communities should look like, and sometimes even violate existing plans.

Planning professor Timothy Duane skillfully illustrates the failings of land use planning in his long-term study of exurban development in the Sierra Nevada. He describes the big stir created by journalist Tom Knudson's Pulitzer Prize–winning series in the *Sacramento Bee*, "Majesty and Tragedy: The Sierra Nevada in Peril," which described how development was degrading the region. Reaction to the series led the state to host a "Sierra Summit," which generated lots of good ideas and enthusiasm for better development patterns. Most of these ideas, however, especially those that would actually affect what owners could do with their land, led nowhere, as development interests effectively blocked new initiatives that appeared to be even the slightest shade of antigrowth. Those initiatives met, Duane concludes, with a "hostile response."[3]

The more telling lesson from Duane's study of the Sierra Nevada is that land use and community planning, even in a state that supports it, is ill-matched to the task of altering development trajectories driven by powerful forces. Duane illustrates in excruciating detail how plans, once adopted, were not implemented; how earlier actions of various commissions and boards constrained later land planning options as conditions changed; how the costs of growth were understated; how buildout numbers and effects were underestimated; and, generally, how pro-development interests politically outmaneuvered growth management advocates.[4]

The promise of land use and community planning is undermined at a few critical junctures:

1. *Getting consent and achieving implementation*: As a democratic process, land use planning and the regulation of development can proceed only with the consent of residents, property owners, and developers in the areas affected. Planners recognize this, and they have put inordinate effort into participatory and collaborative visioning and planning. But there is a disjunction between the feel-good aspects of community visioning, in which residents come together and develop their plan for a healthy, desirable community, and the actual process of land development, in which much of the visioning is either irrelevant or simply ignored by developers, and where elected officials often reject planning staff and planning board recommendations. Planners are complicit in this process in that they go through the motions of visioning, implying that the community can indeed achieve its vision, while their daily job is to enable development that often violates that vision. They smooth off the edges, get design changes applied to development permits, and improve zoning requirements in already heavily developed areas, but they find it hard to achieve the land use patterns that community residents often call for as new areas are developed, such as

building footprints and bulk that fit existing land uses; preservation of agricultural land and wildlife habitat; and a sense of community maintained by broad land use patterns. These weaknesses would reveal themselves more clearly if a stronger monitoring and assessment effort was in place—is development actually following the plan?—and if the public held decision makers accountable to the plan. But, although citizens may take part in the visioning effort, their active participation wanes as the plan is implemented, as actual projects get built, and as the outcomes play out.

2. *Getting a handle on growth*: In most of the West, land use planning is inextricably entwined with growth management. With demographic and economic forces driving rapid regional growth that many citizens view as too much, too fast, public sector planners are naturally under pressure to slow the spread of development, constrain its geography, and otherwise limit growth. Public commitment to growth management is fickle, however, and land use planning is awash in society's ambivalence over growth. We are all at least slightly of two minds about community expansion: bothered by its negative effects on our quality of life, but concerned that the alternative to growth is stagnation and decline. Even community leaders who know better find it hard to see anything but growth—in population, jobs, tax receipts, and, ultimately, land developed—as failure, and they find it difficult fully to believe studies that show that a community cannot grow itself into fiscal balance.

3. *Getting the geography right*: The geography of development and the geography of jurisdiction—of the effective reach of land use planning—simply do not match. Much of the development that is degrading western landscapes occurs in the interstices among municipalities, within broad metropolitan zones, and in the exurbs and rural areas, where land use regulations are lax or nonexistent. Land use planning and regulation are most effective at the local scale and become increasingly weak and vague at broader scales. The development in your backyard, in your viewshed or commutershed, may be in another municipality or another county, and although it affects you directly, you can have no input, nor can your own locally elected officials. We face not only a problem of jurisdiction, but also one of myopia: we all

have trouble seeing the larger picture, the cumulative patterns at the landscape and regional scales, and we find it difficult to conceptualize the collective consequences of creeping sprawl and its multiple effects on everything from traffic to wildlife. If we raise concerns about any particular development, we are accused of a petty, "not in my backyard" mentality, but without help seeing the larger picture, backyard thinking is only natural.

4. *Getting help from the states and the feds*: The West is like two different countries when it comes to the state role in land use planning. The coastal states (California, Oregon, and Washington) play a relatively strong role, with Oregon leading the pack through mandatory urban growth boundaries and detailed statewide planning regulations.[5] In the Interior West, however, state government is of little help in land use planning and regulation, even though the rates and patterns of the region's land development obviously transcend the local sphere. Even the centralized collection of land use data, so vital to assessing the cumulative effects of regional growth, is looked upon suspiciously by many western politicians; thus we do not have the data to assess the status of land use, subdivision, and the footprints of development in adequate geographic detail and at a sufficient scale to make clear arguments for efforts to mitigate effects that swell to the state and regional level. (Some NGOs are building such databases, and are pulling ahead of government in this regard.)[6]

The federal government, even more than the states, is missing in action on the land use planning front. Although federal environmental regulations have some direct effects (e.g., the Endangered Species Act has been the basis for some open space and habitat conservation in the West, most notably in Southern California and southwestern Utah), as do federal transportation, housing, welfare, and defense policies, the federal government has sworn off any explicit role in land use planning. But the emerging land use patterns of the West, especially in among the federal government's huge holdings of public lands, clearly add up in a way that transcends even state interests, especially with regard to water resources, wildlife, and transportation.[7] Worse, several state and federal policies, such as highway expansion and water provision, subsidize and encourage sprawl. If the attention of local decision makers is too often captured by developers promising economic booms and the exigencies of public finance, in which growth appears to be the solution to budget shortages, then it is to the state and federal levels that we must appeal for discipline.

These problems need more attention if we are to devise approaches to overcome them and reap the fuller promise of planning, especially as an expression of a renewed civic engagement in the twenty-first century.

The Conundrum of Participation, Consent, and Implementation

Formal land use planning and growth management are chiefly practiced by local—that is, city and county—planning and zoning commissions, based on citizen input and analysis by professional staff, who make proposals to elected officials for final decisions. For a variety of reasons, local officials do not implement many of the best land use planning proposals.

The planning profession has made an extraordinary effort to involve the public in the planning process, and in many ways it has succeeded: any significant community plan includes dozens of public meetings involving hundreds of participants. But as planner Ken Snyder concludes, although the system has achieved increased public participation, "getting people (even hundreds of diverse people) to show up to a meeting does not necessarily improve the planning process or planning outcomes."[8] Snyder argues that two further improvements are needed in participatory planning: first, it must become more fully democratic; that is, iterative public involvement must work upward from the conceptual stage, where it is mostly stuck today, to the decision-making and monitoring stages. Second, the process should be aided by fuller implementation of communication, visualization, mapping, and impact analysis tools.[9] These two elements must be parallel; that is, the tools by themselves, although much improved recently, cannot create effective, democratic planning.

One big challenge in land use planning is getting durable buy-in from key interests. Planners work hard to get citizen input on plans, but what about consent by developers, and by local officials? Too often citizens achieve agreement on a plan that is only advisory to decision makers. That input is often ignored, or overruled when plans come up for implementation by elected officials, or when specific projects come up for permits—some of which inevitably violate the plan.

Anything short of weakening a community land use plan to the point of ineffectiveness requires that elected officials be fully supportive, and that property owners and developers be fully cognizant of a community's vision and its implications. They must be involved from the start, and the potential outcomes must be so well articulated that there are no surprises when communities adopt and actually enforce plan provisions. Such informed consent[10] is not easy, however, as any plan affects many interested parties, some of which have the ability to thwart it.

Related problems crop up at the implementation and monitoring stages: Is the plan actually implemented on the ground? Can we see its provisions actually reflected in the built landscape? What are the outcomes, in terms of traffic, open space, affordability, and density, and how do they compare to the predictions? More public attention to these later stages, I believe, would make it more likely that the good provisions of plans would eventually be made real. Such attention is especially needed for comprehensive plans that evolve over years, or even decades, as individual development decisions are made. The questions that citizens must ask is whether the cumulative development is actually meeting the goals of the plan, and if not, why not?

Surprisingly few voices have been raised in criticism of the way planning proceeds. First, citizens pour their goals and aspirations into a community vision, and planners turn this vision into good plans, rooted in best planning practices and the expressed desires of the public painstakingly gleaned through countless workshops, surveys, focus groups, and hearings. But then elected officials, at the request of landowners and developers, reject or otherwise evade those plans, often appearing to meet their land use planning obligations without curtailing development potential in any significant way. As Eric Damian Kelley and Barbara Becker argue in their textbook on community planning in the United States, it is often not even clear who has the power to implement plans, and those with the power often choose not to exercise it, especially if the plan actually changes the development potential of land in a significant way.[11]

Actual land use and building permitting too often allows develop-

Figure 8.1 Ridgetop development is often discouraged in master plans; nevertheless, ridgetop homes like this one are common throughout the West. *(William Travis photograph.)*

ment that violates "goals" and "principles" in the comprehensive plan. A classic example in the mountainous West is the many comprehensive plans that call for limiting "ridgetop" development, yet anyone driving through any of the rapidly growing mountain areas can attest to the visible crop of ridgetop homes (fig. 8.1). Kelly and Becker are so frustrated with the process whereby variances from the plan become essentially standard practice that they purposefully write little about the actual development permit granting process because the widespread practice of handing out variances "has little to do with planning and in fact often thwarts the purposes of planning."[12]

The weakness of land use planning is revealed at meetings of county and municipal governments across the West. Planning professor Timothy Duane attended dozens of such meetings in the Sierra Nevada foothill counties he writes about in *Shaping the Sierra*, and I have already recounted his frustrations with a political process that simply thwarts good planning. Bruce Babbitt, an experienced elected official, put the problem bluntly:

> Local governments generally have neither the political will nor the expertise nor the financial resources to stand up to well-financed developers demanding "just one more exception," while lubricating their requests with political contributions. And the occasional local government that does attempt effective planning often loses out, unable to influence what happens just outside the city limits or across the county line, where the jurisdiction with the least environmental regulations often prevails in the competition for jobs and tax revenue.[13]

The gap between community plans and actual development needs to be narrowed. Where can we best shore up the system to make it better at achieving the community values expressed so clearly in so many comprehensive plans? The best approach is to design planning processes that follow through to actual decision making, via a continuing assessment of approved developments and feedback to a committed base of participants who are "kept in the loop," as described in chapter 9.

Growth Management

It is inevitable that land use planning and management will become the locus of growth management, and even growth limits, in most local jurisdictions. With little power and few tools to affect, say, job growth or regional transportation investments, local jurisdictions are pressed by citizens who want slower growth to use land use regulation to decelerate and shape community development, applying their powers of zoning, development permitting, annexation, and land acquisition. It is also inevitable that citizen demand for slower growth will erupt precisely where development forces are most eager to expand construction. This clash of forces not only brings out the ambivalence we all feel about growth, but also yields the pitched battles over land use that are so common in the West. Moreover, the West's "boom-and-bust" mentality works against growth limits. The will to limit development may exist during booms, but there is precious little time and effort available to shape development to serve community interests. During economic downturns, on the other hand, there's no will, but rather the opposite: all efforts go into trying to jump-start growth.

Even in my hometown of Boulder, often touted as an exemplar of forward-looking and effective growth management, there's an underlying fear that maybe, if we're not careful or if we clamp down too hard, our growth management might cause our prosperity to shift into reverse.[14]

A tale of two towns near me reflects the challenge of growth management and the ordeal of places trying to maintain a sense of community as tsunamis of regional growth spill over them. In Erie, Colorado, the battlefront has shifted back and forth across a terrain of pro-growth vs. antigrowth elected officials for almost a decade now; the sides exchange places through both regular and special recall elections. As in so many communities across the West, residents became concerned over Erie's growth after it was submerged by a wave of construction that rushed out along the I-25 corridor north of Denver (Erie was the third-fastest-growing town in Colorado during the 1990s, its population growing 400 percent). In a recent skirmish, citizens elected a slow-growth mayor and mayor pro tem in 2002, who then faced a recall election only eight months later, after they made good on campaign promises to slow annexations and issuance of building permits.[15]

One of the rationales used by the pro-growthers for attempting to throw out the recently elected slow-growth leaders was that the area was booming with commercial development and Erie needed to quickly claim its share of the proceeds. The former (pro-growth) mayor argued that "these delays give other municipalities opportunity to forge ahead with their commercial agendas, diverting sales taxes that could belong to Erie,"[16] and that "we're going to wish to God we didn't have all the houses if we don't move on to commercial development fast,"[17] alluding to the trap set by the town issuing so many residential permits (current annexations and the town plan envision some 10,000 more homes). The pro-growthers didn't feel they could wait two years for another regular election.

The current (slow-growth) mayor retorted that one of the commercial developments she put the brakes on actually planned more residential construction first. The developer had argued that Erie needed *more* houses to support the commercial development that he would eventually build. The frustrated mayor told a *High Country News* reporter that building loads of houses and expecting that "the big box stores . . . will come and pay our way out of debt with their taxes is such a crazy and irresponsible notion."[18]

Next door, one of those big-box operators, Wal-Mart, threatened to leave Lafayette, Colorado, and build across the line in Erie because Lafayette was reluctant to allow it to build a new "supercenter." Lafayette residents had voted in growth limits a year earlier, including an amendment to the city charter requiring the city to enact an urban growth boundary and limits on commercial expansion. The Lafayette town council gave in, undercutting residents' expressed wishes on growth by offering several incentives to keep the retailer because, as a local paper editorialized, "Lafayette couldn't afford to stand by and watch as Wal-Mart took its business—and its large contribution to local sales-tax revenue—across the border to a neighboring town."[19]

The voters had imposed growth limits, according to the editors of Boulder's *Daily Camera*, while "trying to preserve the look and atmosphere of a small town," but they could not control "the growth raging all around them in Erie, Broomfield and other adjacent communities."[20] In Erie, the slow-growth mayor wanted to avoid using big-box retail to solve her community's fiscal problems, hoping instead for businesses that "won't completely violate the idea of a small town."[21] But even she could not simply wave off Wal-Mart.

As the *Daily Camera* editorialized, "It's not easy being Lafayette." Small towns in swelling metro-zones find that, even if the will exists, they simply cannot manage growth alone, without regional cooperation.

Problems of Geography

Any local growth management program faces several hurdles within a community but growth management is especially difficult to pursue at a regional scale. Competition among communities for development, and the inefficient growth that results, is unnecessary and could be curbed by planning tools that operate at a multijurisdictional scale, as well as by changes in, for example, tax policies (e.g., to allow sharing of tax base among localities). A few good and effective regional plans, such as the Sacramento Blueprint and the Puget Sound Regional Council's Vision 2020 + 20 (see chap. 9) notwithstanding, most of the West lacks planning at a scale that effectively addresses the expansive geographic pattern of modern development, both metropolitan and rural. Residents' power to shape their own communities is swept away as an unintended consequence of unguided regional development.

Many of the West's greatest land use problems and threats lie at the lacunae of developed and undeveloped land and in among community jurisdictions, where regulatory zoning ends and only advisory plans (if any) hold sway. In many parts of the West, pro-growth county commissions allow, even encourage, exurban growth in unincorporated areas next to towns. Large-lot development around towns and cities creates awkward geographies of inefficient services and problems as communities grow. Towns find it hard to incorporate exurban landscapes, so towns embedded in counties with few growth controls feel pressed to annex as much land as they can in advance of actual development needs. This pressure, which leads to annexation wars among communities, means that many towns have large undeveloped areas within their boundaries. Just before writing this, I heard the development director for Greeley, Colorado, the nation's fastest-growing town in the early 2000s (in one of the fastest-growing counties), tell precisely this story in a radio interview: the county allowed low-density development on Greeley's edges, which made planning for urban growth awkward, so the city quickened its annexations (creating, unfortunately, some "donut holes" of unincorporated areas surrounded by incorporated Greeley). As a result, Greeley, a town of 80,000, has enough land annexed for 400,000 to 500,000 residents.

On its face, it is obvious that the isolated decisions of dozens of local entities, sometimes in open competition for land, tax base, and jobs, can yield inefficient, undesirable regional land use patterns. Yet regional coordination rarely arises except in cases of large infrastructure, such as beltways and airports and water systems, meant to enable and encourage the spread of development. Multiple governments and agencies cooperated to make sure that Denver had a beltway (as others are now doing to add highways around Phoenix and Salt Lake City), but the local jurisdictions decide, with no concern for regional patterns, how much development goes in at each of the highway's exits. Indeed, they feel pressed to compete, to rush development so as to capture the market for a regional mall, to get their subdivisions going up quickly to signal to the mega-mall companies that the rooftops will be there to support their retail investments.[22]

Partly to overcome fragmented development and partly to stem annexation wars, which sometimes actually slow development, several councils of local governments (COGs) have been created in the West, mostly focused on metro areas where federal highway funding requires that a Metropolitan Planning Organization (MPO) exist to guide transportation investments.[23] The COGs' effectiveness is quite mixed. Although they have power over transportation investment, most have little ability to affect land use; they are better at enabling and accommodating development via highways than they are at shaping it via land use planning.

Beyond their sway over transportation planning, most COGs are merely discussion forums; they may take on issues such as sprawl, but in most cases they manage to achieve only weak, easily revoked voluntary agreements on regional development. They may even be stuck with jurisdictions that miss much of the region's development action. The jurisdiction of the Denver Regional Council of Governments (DRCOG) stops abruptly at the Weld County line, thus excluding precisely that part of the metro-zone now exhibiting the fastest growth.

Fragmentation persists, and the prospect for regional thinking is much brighter than the reality. Well-orchestrated metro-region planning and implementation is under way in a precious few areas; the Sacramento and Puget Sound COGs (see chap. 9) stand out and would make good case studies for those examining that eternal question: what makes some places better than others at planning and managing growth?

Many planners would agree today that regional thinking and planning are in order, but most also assume that actual land use planning authority is destined to remain local, and that even that limited power could erode under pressure from property rights advocates. Planners Christopher Duerksen and Cara Snyder, having examined dozens of cases of regional planning with a focus on habitat protection, are pessimistic:

> Unfortunately, as the case studies so vividly attest, cooperation on regional resource protection is probably at its lowest ebb since the 1970s in this country. Regional government powers and planning are on the wane, and many state programs for local resource protection are slowly being starved by shortsighted state legislatures.[24]

Local governments, they conclude, "actively oppose any type of regional initiatives, viewing them as unholy assaults on local autonomy" (53). Kelly and Becker write that there are "eloquent arguments for regional planning" and a "strong geographical and environmental" logic for supra-local land use planning: "The patterns of development that affect people's lives are regional. Conducting local planning in some ways misses the point." They argue that "only in the context of regional planning can local planning make much sense," but they conclude that the political odds against serious regional planning are "overwhelming" (and they are thinking nationally; what are the odds in the Interior West?). They thus decide to give regionalism little attention in their planning textbook "because it does not exist in this country in any effective sense."[25]

Help from the States and the Feds

As in other areas of public policy, it is natural to look to the state and federal levels for land use frameworks that could transcend local jurisdictions and reduce the negative effects of fragmentation. Professional planners began to push in the 1960s and 1970s for a stronger state role as a way to achieve better coordination and that holy grail of planning: comprehensive plans that capture whole landscapes. State land use planning hit the books with the strengthening of Hawaii's Land Use Law in the 1960s and that state's creation of a statewide plan in 1979. Florida and Vermont followed suit, passing strong planning laws. In the West, Oregon set the pace with its 1973 land use law. California and Washington also provided some state support for coordinated land use planning, but the other western states either ignore the need or actually hinder regional planning (strong home rule legislation works against formal community cooperation in Colorado and Montana, for example). A study of Montana's growth statutes concluded that the state actually limits local jurisdictions' ability to shape growth.[26]

Many professional planners also assumed, right into the mid-1970s, that the federal government would eventually exert some formal land use planning policy meant at least to coordinate state and local processes so as reduce some of the negative consequences of poor development on, for example, water quality or species diversity. The federal government would play this role, they assumed, with national standards and regulations, or with strings attached to federal assistance for highways, urban redevelopment, and water resources. Many planners assumed that U.S. land use policy would eventually "mature" to the European model, with its strong national role in everything from urban design and architecture to countryside protection. The movement almost bore fruit in the 1970s, at the same time some states were taking a bigger role in land use planning. Its chief legislative vehicle was the "Land Use Policy and Planning Assistance Act," passed by the Senate in 1973. The act would have created an Office of Land Use Policy Administration in the Interior Department and would have provided for grants-in-aid to state and local governments, created a national land use database, and coordinated federal and private land use planning.

It didn't work out, as the act fell one vote short in the House. But former Secretary of the Interior Bruce Babbitt argues, in his book *Cities in the Wilderness*, that the federal government is still very much in the business of setting land use patterns, and thus should be in the business of coordinating and planning land use with state and local governments. As Babbitt argues:

> The national government should be concerned with protecting disappearing species, the integrity of river systems that cross state lines, our coastlines, our forests, and regions of special significance for their scenic, ecological, or historic values.[27]

The federal government does have an indisputable stake in the evolution of regional land use patterns in many cases: around national parks such as the Everglades (one of Babbitt's most compelling case studies), along major river systems (which federal activities have extensively modified already, such as the Missouri and the Mississippi, the Columbia and the Colorado), and for resources and natural features of regional and national concern, such as the Chesapeake Bay and the Great Lakes. A larger federal role in land use planning makes obvious sense in several places in the West (e.g., around Yellowstone National Park), but if it does emerge, it will do so slowly.

Can We Plan the West?

Can we ever really "plan" the West? That is, can we compel an outcome different from what unfettered markets and local development decisions would yield? Our land use planning tools, which naturally express communitarian values, are weak in the face of economic incentives for atomized, parcel-by-parcel development; we lack mechanisms for holding decision makers accountable to plans; and our planning institutions are fragmented. But the West is changing, and its citizens are increasingly demanding development patterns that preserve the region's unique attributes as well as their communities' quality of life. Their concerns have been given a voice through the growing debate over "sprawl," and even in a region with a strong culture of property rights and self-conscious antigovernment attitudes, there is an intensifying call for growth management. Citizens wield significant influence over local decisions, and planning, as a democratic process, has the potential, still unmet, to make our visions for better communities real on the ground. Strategies and tools for accomplishing this goal are described in the next chapter.

9 Planning a New West

Strategies for Creating More Desirable Land Use Patterns

It would be one of the real tragedies in the annals of New West land use if Bend [Oregon] simply plods down the pathway of sprawl, ignoring its own potential to become one of the most livable places in the entire eleven-state region.

— Philip Jackson and Robert Kuhlken, *A Rediscovered Frontier*, 2006[1]

THE EVOLUTION of the West's development landscapes—the metro-zones, exurbs, resorts, and the gentrified range—is not haphazard. The patterns are shaped by the logics of development economics, consumer demand and preferences, government investments, and landscape amenities, all of which yield geographic imperatives, such as where the next regional mall would make marketing sense or where housing can be built and sold for the biggest return. As with other dimensions of the economy, a certain efficiency accrues when producer and consumer choices intersect. Furthermore, as with other economic sectors, many of the outcomes of land development are enabled by the public sector. Governments invest in infrastructure, and communities offer incentives to lure more housing and retail businesses. Finally, and unfortunately, as with other aspects of the economy, from health care to agriculture, the coalescence of preferences, markets, and government subsidies does not always yield the best land use from the collective point of view of citizens and communities. Indeed, many of the costs of land development are externalized to the community and the environment. Older suburbs and retail centers are cannibalized to support new developments out on the suburban edge; the values of open space, views, livability, and sense of community (all of which are poorly if at all monetized) are not counted against the raw economic returns from development or the quest by local governments for tax revenue; and the piecemeal approval of individual developments, each seen as having only a small effect, allows larger effects to accrue on the landscape.

Perhaps this is how it should be: the outcome settled by aggressive development forces responding to consumer demand; the developments shaped somewhat by the principles of public sector planning, which are applied with enough force to prevent at least some of the worst outcomes of the unfettered real estate market, yet not applied so robustly as to hamper market efficiencies or invite lawsuits. But in a rapidly growing West, in a region in which communities, even the larger cities, are embedded in a matrix of charismatic landscapes and a fount of naturalness that makes the West wilder than other parts of the contiguous United States, is this the best we can do, as communities and as a society?

Reshaping the American West

Each citizen of the West can take part in shaping its future. The challenge is to strengthen the communitarian forces that can shift development patterns onto better trajectories by employing all the tools of public and private sector land use planning and growth management. Ideas for better development are legion; planning is rich in best practices for land use; and citizens, in public meetings and master plans, routinely call for development patterns that respect the land and community. But our places are less than they can be. If we are to achieve smarter growth patterns in the West, we need to do four big things:

- Integrate land use planning among communities and across landscapes
- Fashion new guidelines for western development, encoding new goals and specific rubrics to yield new land use patterns

- Increase social engagement in land use planning and nurture land use advocacy across the West
- Improve and take greater advantage of land use data, analysis, and simulation tools

We must push the land development complex—private builders, public sector planners, and elected officials—to recognize that land use patterns express who we are as a society, that they reveal what we value, and that they can be altered to create landscapes that better nurture western communities.

Most of the planning tools we need to reshape western growth already exist (box 9.1).[2] Local government has the power, and the legal mandate, to plan land use by establishing regulations about type of use, bulk of buildings, timing and location of development, and even, if it wishes, rates of growth and ultimate buildout numbers. The detailed regulations are guided by comprehensive plans based on community vision and goals, best practices, and the economic, legal, and physical fundamentals that condition land use. The challenge is to apply these tools more effectively.

Integrate Land Use Planning across Landscapes

The call for "regional planning" is almost as old as planning itself, but I don't see western state legislatures requiring (or even encouraging) it in the foreseeable future. The political incentive system is biased toward continued fragmentation of authority on land use and growth. Still, a few voices calling for collaboration across boundaries can be heard, and a few innovative leaders recognize the problems that myopic planning has caused in the rapidly developing West. Denver Mayor John Hickenlooper made regional cooperation a centerpiece of his administration, and when he talks region, he means not only the metro-zone that Denver anchors, but also far-flung rural and resort zones whose well-being is inexorably linked to Denver's. The cofounder of Envision Utah, Robert Grow, also saw the logic of regional cooperation on the urbanizing Wasatch Front, which is so self-evidently connected from one end to the other.

Regionalism has champions in other circles as well. Western scholar Daniel Kemmis not only laid the intellectual groundwork for

Box 9.1

Some Tools for Better Land Use Planning and Growth Management

Tools that operate at the local to landscape scale:

- Detailed comprehensive plans, including specifics on growth, open space, housing, transportation, and natural amenities, linked to detailed land use regulations
- Countywide zoning; zoning districts or planning districts (e.g., rural zoning districts in Montana)
- Site and environmental review of plans for specific developments
- Local open space conservation programs (e.g., Marin County Open Space District)
- Transfer/purchase of development rights programs
- Conservation easements (held by government or a nonprofit land trust)
- Conservation and open space subdivisions
- Special districts (often for services such as roads, water systems, transportation, and recreation)

Tools that operate at the landscape to regional scale:

- Metropolitan regional plans
- Regional planning authorities with LU regulatory power (e.g., California Coastal Commission and the Tahoe Regional Planning Agency)
- Intergovernmental agreements (IGAs) on annexation, development, and open space, often among cities and counties
- State and federal transportation planning
- State lands conservation programs and reforms
- Endangered species Habitat Conservation Plans
- Agricultural land preservation programs (e.g., the Williamson Act and agricultural land mapping program in California)
- Ecoregional and bioregional conservation planning (e.g., the Sonoran Desert Conservation Plan; Greater Yellowstone Coalition; the Wildlands Project; the Yellowstone to Yukon Conservation Initiative; The Nature Conservancy's Conservation by Design)

Planning and land use NGOs:

- Community and neighborhood advocacy groups
- Planning and regional development groups

Box 9.1, continued

- Regional environmental advocacy groups
- Assemblages of land trusts and other local land conservation groups

Analytical aids:

- GIS applications that bring together socioeconomic, demographic, land use, habitat, and biodiversity spatial data
- Urban growth models
- Planning models; community and landscape analysis and simulation tools
- Habitat conservation models

western regionalism, but his Center for the Rocky Mountain West has provided some of the planks for a Rocky Mountain regional infrastructure, such as a regional news roundup, Headwaters News,[3] and a model charter for political cooperation. Arizona's Udall Center and Morrison Institute, California's Great Valley Center, and several other university centers and nongovernmental organizations (NGOs) work on development and land use issues across boundaries, redefining places in more geographically logical ways and providing the analysis and the rationale for regional collaboration.

Of course, it is easier for nonprofits and academic centers to espouse regionalism than it is for local or state officials; in a sense, regional planning is proscribed by planning law. Associations of municipal and county governments in the West have lobbied against statewide planning and legislation requiring regional cooperation (e.g., on annexation) because such approaches reduce local control. Regional planning has been noticeably absent from the agenda of the Western Governors Association (although the WGA has taken up regional transportation, more as a spur to economic development than as a smart growth tool). The federal government, a logical nexus for regionalism, has been little help of late, but perhaps can become, as former Secretary of the Interior Bruce Babbitt suggests, a positive influence for regional planning in the future West. In the meantime, one hopes that federal actions will at least not thwart regional planning by means of, for example, ill-designed energy development, unsustainable uses of federal land, and misaligned transportation planning.

By what mechanisms might we introduce some regional planning into our otherwise fragmented local institutions?

Councils of Governments

In a few places, something approaching regional land use planning has emerged, especially in the West's expanding metro-zones, although seeds of regional cooperation have also appeared in some rural areas, especially around rapidly growing resorts. In the absence of cross-boundary planning, intergovernmental land use agreements are creating de facto regional plans in a few areas. This bottom-up approach to regionalism may well turn into an important basis for new types of planning in the West. It stems from community frustration with what happens "just over the county line."

Despite the mixed review I gave them in chapter 8, councils of governments (COGs) appear to be a viable mechanism for increased regional land use thinking in the West. Dozens of COGs exist in the West, many offering an impressive list of regional goals that resemble smart growth. Not all are effective, however, at shaping growth. What makes some regions better than others at planning and managing growth?

The Association of Bay Area Governments (ABAG), including all nine of the San Francisco Bay Area's counties and 99 of its 101 cities, has an appealing planning process under way: the Smart Growth Strategy/Livability Footprint Project, a "smart growth land use vision" designed to increase the supply of affordable housing, reduce commute times, and improve environmental quality and quality of life in the Bay Area.[4] Yet, as of this writing, there is very little evidence that ABAG's Smart Growth Strategy has yielded any significant planning change or actual land use change in the Bay Area.[5]

On the other hand, 700 miles to the north, around another urbanized bay on the Pacific coast, the Puget Sound Council of Governments has made significant strides toward reforming regional land use patterns. The council has a rocky history: at first no more effective, and by all reports even less inspired, than ABAG, the original Puget Sound Council languished until 1989, when the *Seattle Times* published the "Peirce Report," a seven-part account of the region's growth problems, written by urban analyst and journalist Neal Peirce. The report recommended "purg[ing] the system of the existing more-or-less regional bodies, all of which have run into serious difficulties, none of which now seems fully effective in dealing with the region's problems."[6] Among those bodies targeted for extinction by the report was the "fractious and unfocused" Puget Sound Council of Governments. In 1990, the council reorganized to form the Puget Sound Regional Council (PSRC), focused on growth and transportation issues.[7]

The PSRC includes four counties, 82 cities and towns, three ports, two Indian tribes, and two state agencies. Unlike ABAG, PSRC is blessed with state legislation (Washington's Growth Management Act) that requires comprehensive land use planning and at least advisory urban growth boundaries. In 1990, PSRC developed Vision 2020, a plan that "encourages coordinated and consistent planning throughout each county and the region."[8] The council is now working on Vision 2020 + 20, a strategy that will carry the region through another two decades. The council offers analysis, plans, technical assistance, and cheerleading for regional smart growth, although, like most such regional councils, it has little outright land use authority. It fights an uphill battle against sprawl, but can claim several smart growth successes, especially the establishment and redevelopment of several town centers that concentrate development. A recent study by the University of Washington concluded that such efforts had indeed reduced the overall footprint of growth in the region. This conclusion is supported by the geographic evidence, especially the emergence of denser development and greater continuity of open spaces in several sectors of the region.[9]

What makes the difference between ABAG and PSRC, between successful and less successful regional councils? It's difficult to judge, although several dimensions stand out. First, PSRC works within a context of supportive state legislation that requires local planning and even urban growth boundaries (although quite flexible ones), and it is a locus for federal efforts to protect several species, particularly salmon. Second, PSRC has mustered an impressive array of technical resources for planning in the region, drawing on research, land use simulation, and impact modeling by university and federal government agencies. Effects of growth on open space, wetlands, forests, shorelines, and water quality have all been modeled, mapped, and projected by means that few regions could marshal (again, partly due to the federal requirements). In essence, the Seattle metro area is a laboratory for tracking urban, suburban, and exurban growth and its ecological and social outcomes. Finally, PSRC's planning staff appear to have gained the cooperation of many, if not most, of the towns in the region, both by dint of hard work and because of a willingness to adopt smart growth approaches among leaders and citizens in a region known as environmentally and socially progressive.[10]

Another effective COG is the Truckee Meadows Regional Planning Agency. Although this RPA does not have as much regulatory power as the Tahoe Regional Planning Agency, described in box 9.2, it represents an unusual legislative mandate, by the Nevada state legislature, that two of the state's main cities (Reno and Sparks) coordinate land use plans with the county (Washoe) and work under a collaborative master plan.

A promising example of the COGs' potential influence on land use planning is under way in California's Central Valley. The Sacramento Region Blueprint is the regional development plan for the five-county Sacramento Area Council of Governments (SACOG). It combined a demanding, iterative participatory process with visualization, mapping, and impact assessment tools developed as part of PLACE^{3}S, a suite of tools founded on planning software developed jointly by the energy offices of California, Oregon, and Washington.[11] SACOG planners met three important goals in producing the regional plan: (1) they engaged a larger, more diverse group of participants than in past planning efforts; (2) they included decision makers early on; and (3) they applied an integrated planning simulation tool that was accessible to the participants. The fourth quality of the Blueprint that offers some hope for getting its good ideas implemented (a constant problem, discussed further below) is that PLACE^{3}S is meant to be a permanent assessment tool, kept up to date and available to stakeholders for tracking application of and adjustments in the plan. Of course, this does not necessarily mean that the political will exists to make the implementation phase as participatory as the planning phase or to track and report how implementation diverges from planning, but the promise and mechanism is there.

Planning processes like these are difficult and time consuming.

Still, planners argue that the effort can pay off in plans that are actually implemented, versus those that "sit on the shelf and collect dust" or are otherwise negated by decision makers.

Many other regional councils have at least "vision plans," including the Maricopa Association of Governments' Vision 2025 (the association was formed in 1967 to serve as the regional agency for the Phoenix metropolitan area) and Metro Vision 2030 from the Denver Regional Council of Governments (DRCOG).[12] For the most part, these are weak plans lacking implementation powers. DRCOG extols Metro Vision as being better than plans elsewhere that are "top down and mandatory" because it is "voluntary and flexible" (which means, I am afraid, that it is toothless and ineffective). Still and all, at least such plans explicitly address the landscape and regional scale of development, a vision lacking in much of the West, and they all evince a call for a better quality of landscape and community. Furthermore, they all result from local governments talking to one another about regional issues.

COGs, because they are made up of local elected officials, can have more actual effect on development than NGOs (the best case is effectual COGs supported by NGOs and universities, the situation that appears to make the Puget Sound Regional Council so successful). Although several current western COGs have little purchase on land use and growth management (their mandate is mostly in transportation planning), they are the logical instruments for coordinating growth and land use planning for large functionally, socially, and ecologically linked areas. Who else might coordinate open space systems, water, wildfire hazard issues, and suburban redevelopment across the West's developing mountain fronts, valleys, and plains? In short, westerners should invest in their regional councils. Some existing COGs need to be expanded to include their full metro-zones and associated exurban landscapes, and more non-metro COGs—like the Rural Resort Region that links Colorado's ski towns and their staging communities—should be created in small-town and rural areas now experiencing the kind of growth pressure that often pits one jurisdiction against another. The geographic future of the West will be dominated by enlarging metropolitan areas and associated exurbs, so metro-COGs must be enlarged apace.

Regional Outcomes via Intergovernmental Agreements

Local leaders are often aware of regional problems, and of the fact that their community's land use decisions can cause problems for other communities. Some groups of counties and municipalities hold regular meetings to discuss development issues, and informal consultation across jurisdictional boundaries is as old as planning itself. But few local governments are serious enough about regionalism to overcome the traditions of local control, legal barriers, and other disincentives to engage in formal intergovernmental collaboration.

A chink in the walls that separate towns and counties into land use fiefdoms has started to emerge in the form of intergovernmental agreements (IGAs) on contentious land use issues that lie astride jurisdictions, especially annexation, open space, and adjacent commercial development. IGAs have been used for some time to create agreements (with limited legal force but relatively strong, contract-like social bounds) among local entities. Only recently have they begun to incorporate something akin to the transboundary land use planning needed to achieve broader goals in a politically fragmented landscape.

We have little handle on the number, extent, character, and effectiveness of land use IGAs in the West, although anecdotes indicate they are burgeoning. Here are two examples from California:

- In Yolo County, the cities of Woodland and Davis and the county board of supervisors, attempting to preserve agricultural lands and community separation and identity, signed an IGA to prevent either city from annexing over 11,000 acres of farmland that lies between them. A local reporter noted that "the move, which is not legally binding, is viewed as critical to preserving the distinct identities of the two cities and avoiding the urban sprawl that has plagued many communities throughout the state."[13]
- The cities of Fresno and Clovis signed an agreement that identified where they should grow (in a planning area covering almost 18,000 acres) so as to avoid conflict in services and open space. The agreement specifies growth patterns for twenty years, and is actually a tardy cementing of a joint planning resolution the cities agreed to in 1983, which set voluntary urban growth boundaries that now need to be expanded to accommodate rapid growth.[14]

The Boulder County landscape in which I live is very close to "buildout"—in this case, with significant open spaces remaining—partly due to open space purchases and a remarkably strong county plan that established firm density limits, but also because of a "super-IGA" signed by the cities and the county in 2003, which put in place a binding comprehensive development plan as an overlay to all existing "underlying plans."[15] All annexation proposals are referred to all parties. Of course, the IGA can be amended, and some governments might drop out if they don't like the limits on growth it entails, but as of this writing the IGA is holding. It is not perfect, of course; for example, the area, like many other booming zones, suffers a shortage of affordable housing, and the individual city plans don't add up to sufficient diverse housing stock, so an increasing flow of commuters pours in from the next county, which is a sprawling mess, but offers more affordable housing. Still, these IGAs are examples of some local governments reaching out to shape larger landscapes in cooperative fashion.

The Feds and Regionalism

The role of the federal government in land use planning has been extremely limited. Other than the Tahoe Regional Planning Agency (which was chartered by federal legislation) and the federal Coastal Zone Management Act (see box 9.2), it provides little affirmative support for regional land use planning. In some cases, the habitat conservation requirements of the Endangered Species Act and laws such as the Clean Water and Clean Air acts come to bear on land use patterns, and may even significantly affect local development, but such cases are rare. Yet former Secretary of the Interior Bruce Babbitt identifies several potential roles for the federal government in western land use planning, especially through the Endangered Species Act. He argues that the U.S. Fish and Wildlife Service's Habitat Conservation Plan for the California Gnatcatcher fundamentally changed how housing, roads, and commercial developments were arrayed on San Diego County's landscape. He recounts the shoring up of a ring of federal lands around Las Vegas that will, eventually, rein in that city's spread into the desert. Outside the region, the remarkable, bipartisan, multi-governmental effort to restore the Florida Everglades offers a model

Box 9.2

Federal Spur to Planning for Key Western Landscapes

An Ailing Lake Evokes Regional Planning

Ironically, the regional planning institution said to "wield more power than any other planning agency in the country"[16] is situated in a part of the country that is normally one of the most averse to strong regional planning. The Tahoe Regional Planning Agency (TRPA) acts as the super-planning office for the Tahoe Basin. It has a federal mandate to regulate land use and building pretty much the way any city would, except that its regulations apply to a multijurisdictional landscape and are guided by a core set of planning thresholds, or targets, set by measures of water quality in Lake Tahoe. Based on a 1969 interstate compact ratified by the U.S. Congress (decidedly rare in land use planning), TRPA crosses state, county, and municipal boundaries, has regulatory authority over development in the Tahoe Basin, and can assess fees to cover the impacts of development on elements such as traffic and public services.

TRPA is so unusual among western planning and permitting entities that I considered not including it in this toolkit, as I doubted it could be replicated elsewhere in the West. I was persuaded otherwise after reading Bruce Babbitt's *Cities in the Wilderness* and hearing him speak about the potential when a stronger push for planning comes from the federal government (which he sees as inevitable). Additionally, westerners are seeing more of the region's signature landscapes at risk, and they are looking for bold, perhaps federally aided, solutions.

At Tahoe, according to Senator Harry Reid of Nevada, runaway development, gridlocked interest groups constantly suing one another, and a well-known "national treasure in trouble" finally got TRPA going. These are actually fairly typical conditions in many settings in the western United States (from Yellowstone to the Arizona deserts), so perhaps such planning authorities can emerge elsewhere.

A Beautiful, Fragile Coastline Evokes Regional Planning

In a similar case in which a charismatic landscape and a federal foundation have yielded significant regional land use planning and regulation, the California Coastal Commission operates as a regulatory overlay for development along the coast. California does not require comprehensive growth management as Oregon and Washington do (although it does require community

Box 9.2 continued

growth plans), so the state's strongest land use regulation process occurs under the California Coastal Zone Conservation Act (1976), which was fashioned on the federal Coastal Zone Management Act.

Within the coastal zone, the commission oversees public access to the coast, land subdivision, location and planning of new development, and coastal recreation facilities. The Commission issues development permits in the coastal zone until the local government has adopted an approved Local Coastal Program (LCP), and it then acts as an oversight and appeals board over local governments' permit decisions.

Perhaps more important, the commission limits state investment in infrastructure that attracts or allows development, such as sewage and water treatment facilities and highways (specific protection for Highway 1 preserves it as a winding, narrow road). Of course, where localities can provide funding themselves, this limitation on state investment has less effect. Still, a recent analysis of two counties under intense development pressure found that the Coastal Zone Conservation Act had indeed protected natural resources, reduced agricultural land loss and sprawl, and maintained beach access and aesthetics.

---

that he believes could be applied in the West: for example, in the San Francisco Bay–Sacramento Delta area and in the Columbia-Snake Basin (both of which share with the Everglades a need for reforms in the way that water is managed), as well as in the Greater Yellowstone Ecosystem and the Sonoran Desert, all landscapes in need of repair and larger protection.

Babbitt argues that the federal government has long had the power to affect land use patterns at all scales, for good or bad, although the political will to use this power waxes and wanes. He is confident that it will grow again. What federal roles and tools might we take advantage of when that will again waxes? A simple fact of geography preconfigures the answer: most federal lands are in the West, and the West is roughly half federal land. But federal and local land use planning are like ships passing in the night, hardly aware of one another, despite requirements in both the National Forest Management Act (NFMA) and the Federal Land Policy Management Act (FLPMA) that Forest Service and Bureau of Land Management land use plans coordinate with local land use plans. Secretary Babbitt's optimism that the federal government can, and eventually will, play a more positive role in regional land use management is contagious, and certainly some cross-boundary planning between local and federal lands should be a plank in the West's multi-governmental planning future.

Fashion Land Use Codes for the New West

The efficacy of planning has, for decades, been enhanced by standardized practices and codes meant as national models, often adopted wholesale by local jurisdictions (especially, of course, in urban zoning). Although critics, particularly the New Urbanists, bemoan the one-size-fits-all approach, such national norms help local planners overcome narrow special interests that otherwise would demand land use regulations tailored to their desires. Local decision makers can better hold the line against the particular demands of each interested party by deferring to model codes and "best practices" developed according to national standards that have stood legal tests as well as the important test of time. Model codes are especially useful to local jurisdictions lacking the resources to develop their own land use ordinances from scratch.

Standards for land use are well developed for zoning, design, construction, and various health and safety matters, but are less available for landscape-scale themes such as comprehensive land use planning, open space, habitat protection, and the many, subtle dimensions of the overall shape and look of communities. But the seeds of land use standards suited to landscapes and tension zones in the West are starting to appear, and they deserve our attention.

Western Guidelines

The Lincoln Institute of Land Policy and the Sonoran Institute concluded that, despite urban infill and redevelopment efforts, more than half of the new growth in the West over the next quarter century will be greenfield projects at the suburban edge. So the two groups analyzed a dozen cases of smart urban-edge development and collected the lessons in an easily accessible report, "Growing Smarter at the Edge."[17] They found that the best master-planned communities encompass several basic smart growth elements, including integrated open space; mixed public, commercial, and residential uses; pedestrian orientations and alternative transit; and a range of housing densities and prices. But the report concludes that the benefits of "smart growth at the edge" accrue only if local jurisdictions have detailed master plans in place and enforce them. Detailed guidelines for master-planned communities are also important, as is a community's commitment not to "budge on the basics," such as open space and a mix of densities and prices. The report concludes that well-designed urban-edge developments can reduce the negative effects of growth on western landscapes.

The challenge of developing and applying smart growth standards out past the suburban edge, in rural areas and around small towns, is heightened by weaker planning infrastructures: little or no zoning and only weak, often vague master plans (if any at all). Planners Christopher Duerksen and James van Hemert, in their book *True West: Authentic Development Patterns for Small Towns and Rural Areas*, distill a code for western rural development from case studies and strong doses of history and natural landscape design.[18] They argue that most contemporary notions of small-town and rural development are based in eastern or midwestern places and simply don't work in a region of dry climate, wide-open spaces, and majestic scenery. They assemble 83 rural development guidelines, ranging from the landscape to the streetscape to the site scale. At the landscape scale, their key recommendation is to maintain, and guide development into, existing towns while preserving open spaces that separate communities. They also recommend developing new, concentrated towns rather than allowing existing communities to grow too big (they suggest that small towns choose an absolute limit on their size, certainly no larger than Mormon leader Joseph Smith's 15,000–20,000 village concept, and they lean more toward examples with a population of 5,000 at buildout). Reckoning that dispersed rural settlement is also inevitable, they lay out guidelines for houses, roads, and other improvements meant to reduce its ecological and visual footprint. Attention to public places, sacred spaces, views, agriculture, water conservation, and wildlife habitat round out their rural template. Duerksen and van Hemert are refreshingly candid in stating that these development patterns will rely on detailed master plans, ordinances, zoning, and overlay zones—in short, on regulation. That such regulation can be effective, and can enjoy community support, is confirmed by their many case studies.

True West reflects the cultural traditions and ecological patterns of the Rocky Mountains and the Southwest, and some of the principles may be different in the Canyonlands, Central Valley, and on the great Columbia Plain. Fortunately, some bioregional and ecosystem groups are working with planners and designers to develop rural development guidelines that suit their subregions. For example, the Greater Yellowstone Coalition is developing a detailed guide to rural residential development for the Greater Yellowstone Ecosystem, stressing several key regional problems such as wildfire, the need to protect streamside habitat, and the particular problems of rural development in a region that hosts large herds of migratory wildlife as well as the predators that follow them. We need other similar groups to develop "best practices" for their ecoregions.

Community Character

The recommendations in *True West* rely on a shared sense of community character, which can be a challenge to articulate in a region swelling with migrants from elsewhere. Yet one of the defining features of the long history of human settlement has been the emergence of geographic communities, places that become the focus of residents' sense of belonging, security, and identity, *even among newcomers*. Today, even in the sprawls that our cities have become, people identify subsets of the city as their place, their community, their neighborhood, their shopping district. When asked to draw a map of their communities, residents of large cities often draw their neighborhoods, delineated by certain well-known but unofficial boundaries, such as a road or a park or perhaps a more subtle change in quality of the housing stock, and often centered not on their home but on a civic facility such as a library or park. I have met many residents of small towns and rural areas who counted not only the whole town, but an entire valley, as their community, inverting the urban dwellers' search for a more constrained place.

Planners have long recognized this sense of place and community, and planning practice routinely seeks to encourage community identity and the patterns of place that anchor it. But development and jurisdictional patterns, and too often, zoning and development codes do not comprehend the geography or character of community. Ongoing battles over the super-sizing of homes in older neighborhoods of Seattle, Denver, and Boise all attest to this problem, as do arguments over fencing, gated communities, and house size in gentrifying rural areas and resorts.

The gratifying planning response to community concerns has been to try to codify community or neighborhood character in metrics such as building square footage, footprint, bulk, landscaping, setback, and profile. But, as every planner knows, such metrics miss much of the "stuff of community." A broader field of community indicators has been developed to help people track, and track down, their places.[19] Yet the indicators approach is too focused on social measures such as education, crime, child welfare, and even counts of derelict buildings and cars, and not sufficiently well developed in terms of land use indicators that speak to community character, such as open space or trails accessibility, time to buildout, rate of development versus rarity of local habitats, and the mix of income sources that constitute the local economic ecology. Careful attention to such measures may yield surprises, as when a resort town discovers that it is not as dependent on tourism as it thinks, or that even modest developments in the wrong location can eat up critical habitat faster than expected.

Another challenge is to get to the essence of community identity, to discover, as it were, the "heart and soul" of a community. Perhaps the problem is more along the lines of getting to realistic and actionable elements of community character. Although community vision statements might allude to economic sustainability, diversity, rural character, and so forth, the more demanding measures might be, for example, jobs that pay a living wage (in relation to local costs of living); the proportion of resident to nonresident homeowners; the proportion of the workforce housed in the community; and the mix of services available (resort towns might lack typical services such as cleaners and clothing stores, whereas whole poorer urban neighborhoods might lack a decent grocery store or gym).

Box 9.3

Guiding Principles for Seattle's Open Space Plans

These guiding principles, taken from "Open Space Seattle 2100," a coalition of interests formulating plans for Seattle's integrated open space program, define best practices for the social, ecological, and economic utility of open space protection.

1. Regional Responsiveness

Consider Seattle's role as an ecological, economic, and cultural crossroads; its location in one of the world's great estuaries and between two dramatic mountain ranges; its critical position as a threshold to two major watersheds (Cedar and Green/Duwamish); and its relationship to salt and fresh water bodies throughout the city.

2. Integrated and Multi-functional

Integrate a variety of types of open space within a unifying, coherent structure. Incorporate considerations for streets, creeks, parks, habitat, urban forests, trails, drainage, shorelines, views, commercial and civic spaces, back yards and buildings. Consider layering multiple functions and uses within green spaces to create high-functioning, high value open spaces.

3. Equity and Accessibility

Within a network of open spaces provide equitable access for all persons to a variety of outdoor and recreational experiences. Distribute appropriate open space types to every neighborhood, in order to address the needs of diverse population groups. Prioritize public access to water.

4. Connectivity/Coherence

Create a wholly connected system that facilitates non-motorized movement, enhances habitat through connectivity, links diverse neighborhoods, and is easy to navigate and understand. Connect these in-city amenities to surrounding communities, trails and public lands.

Box 9.3 continued

5. Quality, Beauty, Identity and Rootedness

Use Seattle's many natural strengths to create an exemplary, signature open space system. Build on intrinsic qualities, both natural and cultural; reflect, respond to and interpret geographic, ecological, aesthetic and cultural contexts; address emotional and spiritual needs; and inspire a deep connection to place.

6. Ecological Function and Integrity

Expand the quantity and quality of natural systems in the city: Provide quality habitat for all appropriate species, with a special emphasis on the waters' edge. Design for hydrological health (water temperature, water quality, water regimes, stormwater), and consider appropriate water and resource conservation strategies. Connect to regional ecosystems in order to achieve integrity, resiliency and biodiversity in ecological systems in the face of climate change.

7. Health and Safety

Continue to make the city a safe and healthful place to live. Reduce the risk of natural hazards (slides, flooding, earthquake, soil and water contamination) while reclaiming and treating previously toxic sites. Provide multiple opportunities for exercise, physical activity, and a connection to nature to be integrated into daily lives.

8. Feasibility, Flexibility and Stewardship

While visionary, the plan should be lasting and feasible, with a complementary set of near-term implementation strategies that includes mechanisms for both public and private investment that are achievable in incremental steps and adaptable over time (e.g. codes, funding sources and incentives). It should be maintainable, inspiring shared stewardship between public agencies, private businesses, and individual citizens to foster pride, purpose and community.

Source: Open Space Seattle 2100 "Plan Goal Guiding Principles, Process Goals." Final Draft. (Department of Landscape Architecture, University of Washington, Seattle, 2005) 1–2. Available at http://depts.washington.edu/open2100/ (accessed December 6, 2006).

The challenges, then, in preserving community character are threefold, and they apply to communities in all settings in the West:

- Develop agreed-upon measures of community character
- Codify desirable goals in measurable and actionable terms
- Implement those goals in local planning and growth management policies, and then track the results

Coding Green Infrastructure into the New West

Model land use codes for most western locales and ecosystems will include provisions and guidelines for maintaining open space, which I believe will become the single most important ingredient of regionally appropriate development. Open space will serve both social and ecological needs; it is the prime element of the region's essential green infrastructure.

Strategic planning for open space should consider broad guiding principles and the many roles that open space plays in a region's natural and social ecology, as Seattle's planning process does (box 9.3). More detailed rubrics might include how representative a protected area is of local habitats, whether it has historical significance, and even production values such as agriculture. The challenge is to make the rubrics sufficiently detailed: for example, what if a planning principle guaranteed every resident, from Seattle or Tucson or Helena, access, within walking or biking distance or a bus trip from their homes, to trails that would allow them to walk to public lands and then, essentially, to any other part of the West? It does not matter that no one would ever walk with a backpack from a bus stop in southwestern Denver to, say, Moab, only that the network of trails and open spaces would be in place for someone to do exactly that, providing a variety of services and benefits, from social to ecological, even for persons who never physically use those spaces.

This logic is already afoot in the West in the town-to-mountain trail systems created in Bozeman and Boise[20] and in the Pleistocene Trail System along the benches that demarcate the eastern border of Salt Lake City. But these efforts remain piecemeal (although locally quite vital) land protections. They need to take their place in a matrix of protected open spaces that add up to more than the sum of the parts. Yes, Boise should protect its mountain scenery, which, as the Boise Parks and Recreation Department points out, "provides a postcard backdrop that inspires and soothes the soul." But the Boise Foothills should be a piece in a larger puzzle that, for example, links the Snake River Plain with the mountains.

How do we set priorities for protecting open space? The most obvious, and inevitable, priority is set by development itself. Land threatened by development, which is almost always land near expanding suburban and exurban landscapes in the West, must have priority protection. The Boise Foothills protection program emerged simply because the growth of the metro area threatened to leap right up the foothills, into that "backdrop that inspires and soothes the soul." When applied in this way, land protection not only protects land, but also builds human relationships with one another and with the land.

A second priority must be the needs of nature. We might best use the term "natural spaces" rather than "open space" here because "open space" is simply too vague, applying to golf courses and heavily cropped farmlands as well as to natural landscapes less transformed by human action. Most natural spaces play the role of open space, providing recreation, views, and community buffers, but they also supply critical ecological services: sequestering carbon, trapping floodwaters, cleansing runoff. Providing space for nature is challenging, however. Nature is a demanding taskmaster; it wants room for wildfires to expand, floods to spread, and species—some of which don't get along well with humans—to move. These natural phenomena require different sizes and configurations of open land. In the spirit of choosing a limited number of goals that can then act as an umbrella protecting other landscape qualities, one compelling goal would be to identify important wildlife migrations in the West, both local and regional, and protect the natural spaces needed to maintain those migrations. Migration, of course, requires core home territories and different types of seasonal habitat, but the migration routes themselves are an indisputably necessary ingredient that touches multiple types of landscapes and thus obliges us to protect lands in areas and configurations that we might not otherwise (e.g., our tendency would be to protect core habitats alone, as many of them are already on public lands, as well as open spaces that mostly serve human needs). In addition, wildlife migration corridors naturally cut across political jurisdictions and thus demand interjurisdictional collaboration.[21]

Another obvious focus for protection of natural spaces is stream corridors. The West is dry country, and every running stream and river is a treasure. But by virtue of history and geography, streams and rivers have attracted development and transportation infrastructure, and they are ecologically stressed. Much of the riparian landscape of the West is private, at the lower elevations where homesteaders first settled, so few stream miles are protected from development. Yet these important landscape features provide social and ecological resources out of proportion to their geographic extent: they are vulnerable, important, and rare landscapes; they deserve priority protection.

Finally, green infrastructure must also include big thinking and big plans. Take the Mountains to Sound Greenway Trust, formed in 1990 to turn the I-90 corridor, a fully 100-mile swath that bisects the Cascade Range from Seattle to Thorp, into a protected zone. Outside of existing urban areas, the trust has worked to protect private parcels, coordinate land use and habitat protection within the extensive public lands that lie astride the highway, and develop a network of trails that make the greenway more accessible to the public.[22] The Sonoran Desert Conservation Plan, a collaborative effort of Pima County, Arizona, and dozens of NGOs and agencies, covers some 5.9 million acres and encompasses over two dozen threatened and endangered species.[23] Even more breathtaking is the Yellowstone to Yukon Conservation Initiative, which doesn't hesitate to plan for a corridor fully a thousand miles long across several states and two nations.[24]

Of course, the green infrastructure of the West naturally includes its federal and state lands, which, in their domination of the West's rural geography, are unique in the nation. So much has been written about, and so much effort has been expended on protecting, the federal lands that little need be added here. The state lands across the West represent a less recognized fount of open space and habitat (constituting some 146 million acres), and they are more threatened by development because most state lands can, if their governing commissions choose, be sold for private development. State land reform campaigns in Colorado and Arizona have blunted this risk somewhat in those states, especially in Arizona, where extensive state lands adjacent to Phoenix and Tucson were in the path of those cities' rapid sprawl. Fortunately, state lands protection campaigns are springing up across the West.[25]

Ultimately, some form of comprehensive ecological planning is needed across the West, across all land categories; but experience and research tell us that integrating ecological dimensions into comprehensive planning will be difficult and controversial. So a program of tactical preservation, on both private and public land, is in order, one that includes every open space effort, no matter how small, in a broader scheme of ecosystem and regional "conservation by design" that helps us value the ecological and social role of each open parcel.

Make Public Sector Planning Work for Communities

We must increase public engagement in land use planning and nurture land use planning advocacy across the West. This effort will fundamentally strengthen the public sector's ability to shape land development so that it reflects a community's goals. In many cases, this simply means pushing our elected officials into paying attention to existing plans, most of which already reflect a community's values, and actually implementing those plans with tools already at their disposal. But the continuing gap between plan and implementation indicates that something more is needed to achieve community goals.

That gap is largest outside of municipalities, where planning and zoning are weak or nonexistent. Although many areas have countywide "zoning," the term hardly lives up to its connotation. Almost half of Wyoming's counties have countywide zoning in place, for example, but the zoning is weak, it is typically of limited value to land conservation, and it may even encourage sprawl by requiring orthodox, large-lot rural residential development. Still, the county comprehensive plans that provide the basis for this admittedly weak zoning are worth the effort, and every county in the West should have a comprehensive land use plan in place (as required by most western state laws). Where practiced, county planning is becoming more comprehensive, at least conceptually, addressing issues not typically found in plans up through the 1980s. Most plans now include housing, transportation, open space, community separators, wildlife habitat, slope protection, water and air quality protection, and agricultural land protection. The plans may lack regulatory teeth, but they are coming more and more to reflect, at least loosely, smart growth principles. Comprehensive plans include a community's vision for its future, and without exception that vision is a positive one that stresses quality of life. Once in place, these plans can provide the basis for residents and planning advocates to push elected officials in the right direction. Planning alone won't change the face of the West, but it is the foundation for change.

Make Comprehensive Plans Enforceable

In something approaching a cynical manipulation of social capital, the forces arrayed for business-as-usual growth are happy to see community effort poured into master plans with little political power and few land use teeth. Professional planners recognize this problem and have pushed state and local governments across the West to make comprehensive plans more legally binding. Recommendations by different planning organizations vary, but generally include state legislation that allows, but does not require, each jurisdiction to make its comprehensive plan binding on land use and development; that is, to make plans more binding on local officials rather than only "advisory," the gray area in which they commonly reside.[26]

Now it may happen that as plans take on a more decisive role in actual land use outcomes, they will be watered down. But they will also become more definitive and worthwhile, and suddenly, actors who do not bother to take part in the comprehensive planning process will be compelled into the room. The decisions that really determine how a community will evolve and grow will take place more often in a planning setting and less often in the make-or-break urgency of town council and county commission meetings taking up individual development projects. The whole community dynamic will be changed, and a new social contract will be forged: developers and property owners who could in the past avoid face-to-face discussions with other community residents will have to bring their goals and plans to the more foundational process of community visioning. This already happens in communities that take their master plans seriously: jurisdictions with a track record of implementing their plans find that various interests more effectively involve themselves in the planning, not only in the deciding or the protesting.

Create a Watchdog for Every Plan

A more robust planning advocacy must accompany a more robust planning process. We need to nurture the "quality of life lobby" across the West. Every comprehensive plan in every jurisdiction in the West should have a standing "watchdog" group pushing for, and monitoring, its implementation and its improvement and regular revision.

Where it is impossible to have a standing group dedicated to each and every plan, a regional watchdog group should be put in place and should take responsibility for a land use and planning watchdog circuit, tracking comprehensive plan updates, development permits, and other land use actions in selected towns and counties. Land use activism can (and often does) share people, resources, and issues with environmental and social advocacy, but the particular institutions of land use decision making deserve unique attention from dedicated interests with land use expertise.

One obvious rubric for organizing such land use watchdogs is bioregional. It is an overused term, but grassroots efforts around land and water issues that transcend political boundaries and are instead aligned with natural features and boundaries are sprouting up and gaining credibility and power in the West. A leading framework for this effort has been the watershed, and hundreds of watershed-based initiatives and collaborations, some with land use interests, are under way in the West. A second approach has been to delineate an ecosystem, like the Greater Yellowstone, or an even larger, connected swath, like Yellowstone to Yukon or the Cascades, and array advocacy around threats to that landscape.[27]

Unfortunately, these bioregional advocates have tended to neglect community development and land use issues, focusing instead on protecting nature. Still, their attention to recognizable landscapes and definable ecosystems is a potentially powerful framework for land use advocacy, and some groups have specifically taken on local land use planning as a tool for achieving ecological and social sustainability: the

Greater Yellowstone Coalition and the Grand Canyon Trust have added programs on community development and regional smart growth to their main missions of federal lands advocacy.

Land use plans are too important to the West's communities to be the subject of only temporary and informal coalitions of neighbors and other advocates that form essentially on a development permit-by-permit basis. Nor can we count on other special interest groups to carry the burden of better community planning. Yes, environmental, housing, and social justice groups do step in, but their expertise, and their attention, is not always in land use and growth management. We need, first, to build a roster of standing land use watchdog groups in the West, and we need to develop a recruitment and support program to seed such groups where they are needed, in different settings (metro to rural; desert to mountain) and at different scales (municipal, county, and regional). The most important investment is funding to employ a planning leader in the communities because 90 percent of planning advocacy is "showing up"—at every board meeting, every planning commission study session, every focus group. Next, we need to support these advocates with training and data because showing up informed and armed with data (maybe more and better data than the commissions themselves have) adds power to the traditionally weak planning forces in our communities.

Land use NGOs may also take on the role of government watchdog groups. The leading example, 1000 Friends of Oregon, was formed specifically to monitor the implementation of Oregon's state growth management law. PLAN-Boulder County (Colorado) formed in 1959 to advocate for a growth boundary on the slopes above Boulder (the "Blue Line" above which city water would not be extended) and was instrumental in creating Boulder's open space and growth management systems.[28] A few other "1000 Friends" groups have developed, even in places, such as New Mexico, with no statewide planning, where they act as smart growth advocates.[29]

Land Use NGOs: New Kids on the Block

Neighborhood and community activism on development and land use has been around for a long time, but it has one major flaw: it waxes and wanes with each land use battle, and it is firmly entrenched in classic NIMBY (Not In My Backyard) politics. Environmental advocacy groups, too, have been around for a long time. The focus on land use per se is newer among NGOs, as is the scale at which they operate.

Four main types of land use–oriented NGOs have emerged in the West: community and neighborhood advocacy groups, good planning and regional development groups, regional environmental advocacy groups, and assemblages of land trusts and other local land conservation groups.

Community and Neighborhood Advocacy Groups There is nothing fundamentally wrong with the ebb-and-flow pattern of neighborhood and community activism. Neighborhood groups ought to form as needed to protect their particular places, to seek equity and democracy in decision making, and to get the attention of bureaucracies often ill tuned to the grassroots. It is precisely when significant land use and development plans are in play that emergent neighborhood groups can have the most leverage. Furthermore, the fundamental issues they often raise remain roughly the same. Though each project may differ, local groups often find themselves pursuing similar overall goals:

- Development should meet local needs and fit the neighborhood milieu.
- The planning process must be inclusive and democratic.
- Planning actions must address equity (e.g., cost of housing, equitable access to services).

Still, the challenges of smart growth and sustainable development, from the locale to the region, are best met with enduring engagement. Planning professor Timothy Beatley argues that community engagement with land use must also have a more positive attitude:

> Long-term commitment to sustainable places will require a politics in which people and organizations work together to create a positive future, not simply to oppose specific projects or decisions they deem threatening.[30]

Beatley and others think in terms of shifting the process from fighting projects to advocating for positive developments in every neighborhood and community ("Yes-In-My-Backyard," or YIMBY). Many urban planning institutions now formally recognize neighborhood groups, saving, as it were, a permanent place at the table for them. A mature place-based advocacy would bring together multiple threads that have long been part of comprehensive planning: appropriate design, efficiency of use and location, access to services, affordability, and environmental protection. Planning scholar and activist William Shutkin makes the connection between neighborhood advocacy and environmentalism, arguing that environmentalism must be as much about protecting local, even urban, places as it is about protecting wilderness areas or the global climate.[31] And, of course, the neighborhood development and restoration projects that he and Beatley describe, such as transit-oriented redevelopment in impoverished enclaves of Oakland, California, or brownfields reuse in Tucson, Arizona, are the building blocks for solving larger environmental problems and reducing social inequities. Recent arguments over the siting of affordable housing in Vail, Colorado, illustrate that such struggles are not limited to urban areas, or to poor neighborhoods.

Enduring local planning advocacy will require nurturing by local governments, but also by the regional planning NGOs.

Good Planning and Regional Development Groups Local battles over land use and development can lead to lasting efforts. The Sonoran Institute got its start in 1991 because of development tensions between Tucson and the nearby Saguaro National Monument. Now the Sonoran Institute is the leading example of what might be called "good planning" or smart growth advocacy in the West. Such organizations are tough to categorize, their mission statements reflecting the ineluctable goals of sustainable development. The Sonoran Institute, for example, expounds what it calls "community stewardship":

> The Sonoran Institute's community stewardship work creates lasting benefits including healthy landscapes, vibrant economies, and livable communities that embrace conservation as an integral element of their economies and quality of life.[32]

"Conservation" is writ large in the Sonoran's mission and programs, but it is never very far from "development."

Most "good planning" and smart growth NGOs in the West focus on a particular region, either metropolitan or rural: Envision Utah on that state's metropolitan Wasatch Front; the Cascade Agenda on the Central Cascades and Puget Sound; and the Great Valley Center on California's Central Valley. They pursue various mixtures of community development and environmental goals. The Great Valley Center, with a focus on agriculture, community development, and environmental issues such as air pollution, finds itself inevitably drawn into the land use and planning issues that are at the root of so much of what makes for community and environmental sustainability. Like many such NGOs, it catches people's attention by developing scenarios of future development, much like the American Farmland Trust did for the same area (and Envision Utah for the Wasatch Front). In some ways, NGOs are freer to map out future land use over large areas than are planning agencies, which operate in an often more tense regulatory environment in which "lines on maps" incite concerns of property owners.

The Orton Family Foundation (which provided support for this book) seeks to improve community and land use planning practice in New England and the West. The foundation sees land use planning as the "pathway to vibrant and sustainable communities." But it also acknowledges the challenges of land use planning in its mission statement: "Land use planning, one of our most important civic responsibilities, has not lived up to its full potential." Indeed, like Sonoran, Orton is pursuing social change, and they both recognize that the process of planning, not only its outcomes, matters.

It is still not clear what community sustainability would look like in the West. Certainly it would include both ecological and social components. The Sonoran Institute throws in a healthy dose of "New West School" economics. It deploys data and analysis to support the argument that extractive economies were unsustainable and that they squandered the very resources—natural areas, open space, water, fish and wildlife—most valuable to community development in the postindustrial era. It works to give local officials data to assess these values and ways to incorporate them into community plans. The Orton Family Foundation focuses more on the process of planning, working on ways to make it more participatory, more democratic, and more fruitful, especially, for instance, through simulation tools and scenario planning approaches.

By making a commitment to both environment and community, and to growth and land use management tools that include both incentives and regulations, the Sonoran Institute appears to have made inroads within the antiplanning Interior West. It has gained traction especially with county commissioners around the West through the workshops given by its Western Community Stewardship Program (in partnership with the National Association of Counties). One recent assessment of these county forums describes local officials coming out of the workshops and deploying both traditional and innovative tactics, ranging from rural zoning (Custer County, Colorado) to agricultural land protection ordinances (Rio Arriba County, New Mexico) to the purchase of development rights (Sublette County, Wyoming).[33] Still, some planning and growth management efforts at least partly spurred by Sonoran's stewardship forums have run afoul of pro-growth, antiplanning, and pro–property rights attitudes. At this writing, the Sublette County program to buy development rights to private land designated as open space is limping. Park County, Montana, tried a buildout study and a revamp of its comprehensive plan after participating in a 2003 Sonoran forum (and because the Greater Yellowstone Coalition sued to force it to comply with state laws requiring comprehensive planning), only to have the plan turned down in December 2004, when "hundreds of people showed up at a public meeting and angrily denounced it." The county proposed a new growth management policy in November 2005, which stresses property rights and voluntary efforts to protect agricultural land. It includes few hard recommendations on land use and offers no proposed regulations.[34] Even with "good planning" NGOs expanding across the West, the road to smarter growth is long and winding, especially in the interior.

A business core marks several other regional NGOs with land use interests, such as the Sierra Business Council (SBC). SBC emerged from growing concerns over the degradation of the Sierra Nevada, concerns propelled by journalist Tom Knudson's 1991 series of articles on the range in the *Sacramento Bee*. It was purposefully organized as a coalition of business owners who counted on the quality of life and the environment in the Sierra for their livelihood (not only tourism businesses, but also wood products companies and even builders who supply housing to the region's rapidly growing population). The SBC broke new ground in the region by attempting to measure both natural and social capital, as well as the financial health of the region, via the "Sierra Nevada Wealth Index," a business approach that defined, in a sense, the regional "bottom line," a balance among extraction, development, and preservation.[35] SBC is also remarkable for having developed early on a community and land use planning program as a formal part of its services to towns and businesses in the Sierra Nevada region.[36]

Regional groups can also call on national organizations for help with their smart growth agendas. The American Planning Association, the Congress for the New Urbanism, and, especially in terms of regional land use, the Growth Management Leadership Alliance[37] now offer guidelines and establish professional standards for better land use.

Regional Environmental Groups One of the most important developments in western environmental advocacy has been the emergence of several influential groups tied to specific landscapes, ecosystems, or subregions. At their core, these groups are mostly traditional environmental advocacy groups, formed at least partly to fill the regional gap left by national organizations. In some cases, a regional group formed because no national group could focus sufficient attention on a specific area; for example, the Southern Utah Wilderness Alliance formed as the key regional proponent for wilderness protection in Utah's canyonlands. These regional groups are, at their core, environmental watchdogs, fighting for species protection and monitoring forestry, grazing, and mining on federal lands. But as they have developed more comprehensive advocacy for regions and ecosystems, they have also found themselves drawn to private land use issues.

The Greater Yellowstone Coalition formed a community development program to focus on land use and smart growth in the twenty-two counties that touch on the Greater Yellowstone Ecosystem. This effort, never as sizeable as the coalition's work on federal land and species policies, nevertheless brought the group into a new realm of activism, shifting its attention from environmental impact statements and the management plans of a few federal agencies to the activities of scores of planning commissions, special districts, open space commissions, county commissions, and town councils. The coalition collected data on comprehensive plans, subdivision rates, and population growth; conducted studies of the cost of rural services; hosted a national conference on applying smart growth principles to the rural areas and towns around nature preserves; and occasionally found itself protesting specific development proposals.[38] This attention logically led it to take a prescriptive approach, working with local planners and open space groups to develop something akin to "best development practices" suited to the environment it protected.[39]

Many other ecoregional groups and programs in the West, such as the Southern Rockies Ecosystem Project and the Yellowstone to Yukon Conservation Initiative, have found themselves drawn into local and regional land use planning advocacy as they come to realize that the goals of habitat and species protection and landscape connectivity cannot be achieved by the traditional focus on public lands.[40]

Land Conservation Groups The rise of land trusts and other NGOs focused on private land protection is the most striking and important innovation in open space and habitat conservation in decades (fig. 9.1). Land trusts purchase land or development rights from landowners and hold them for conservation purposes. Aided by growing philanthropy for land conservation and by state and federal laws that offer incentives to private landowners, the land trust movement is reshaping private land use, especially in rural areas, often in ways that local planning could not.

Because buying land outright is more expensive, the trusts work mostly through conservation easements: legally binding agreements, enshrined in land titles, whereby a property owner relinquishes the right to develop land, often in return for reduced taxes. In his recent study of conservation easements, lawyer Jeff Pidot writes:

> No recent happening in land conservation rivals the rapid deployment from coast to coast of conservation easements. Beyond tax and other public subsides, a driving force fueling this phenomenon is the perception that conservation easements are a win-win strategy in land protection, by which willing landowners work with private land trusts or government agencies to provide lasting protection of the landscape.[41]

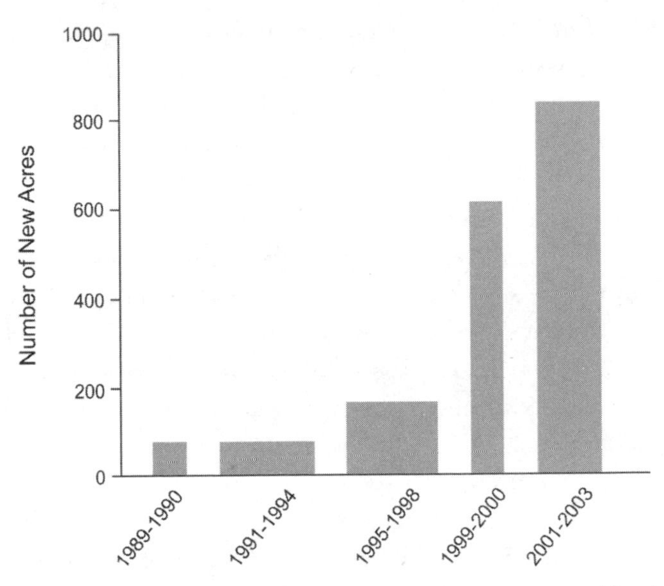

Figure 9.1 Numbers of new acres protected by local and regional land trusts between 1989 and 2003. *(Modified from Jeff Pidot, "Reinventing Conservation Easements: A Critical Examination and Ideas for Reform," Policy Focus Report 13, Cambridge, MA: Lincoln Institute of Land Policy, 2005, 1. Data from the Land Trust Alliance.)*

The rapid growth of local and regional land trusts reveals an unmet demand for private land conservation. The Land Trust Alliance identified 1,500 trusts in the United States in 2003,[42] but no central database exists to assess how much land, in what geographic settings, and with what restrictions, is protected by the trusts. Most land trusts are truly local; many came into being to protect a specific property. The Land Trust Alliance has developed standards and practices for trust activities, including requirements for strategic planning and land and environmental assessments.[43] The movement is so fragmented, however, that we cannot yet judge its broader land use and environmental outcomes. Most trusts respond to opportunities rather than following a broader blueprint, and the types of land and protection purposes vary greatly, from urban parks to working farms and ranches.

A few state and regional land trusts in the West do manage to work strategically. The Montana Land Reliance, for example, operates like most land trusts, protecting everything from historic homesteads to particular wetlands. It also invests heavily in two threatened ecoregions: the Northern Continental Divide Ecosystem and the Greater Yellowstone Ecosystem.[44] Other trusts operate across large areas, but with specific focuses: ranch and rangeland trusts have formed to protect working ranches in the West, such as the California Rangeland Trust and the Colorado Cattlemens' Agricultural Land Trust.[45] In some cases, an umbrella organization has emerged to assess regional land conservation progress, raise funds, and set goals based on needs assessments. The Colorado Conservation Trust, for example, acts as a strategic assessment, planning, and fundraising organization positioned to assist local land conservation groups. Its 2005 statewide assessment of land conservation successes and needs identified forty-six land trusts and forty local open space programs that had protected a total of 1.6 million acres (through a combination of outright purchase and conservation easements), but its needs assessment identified 2 million acres in need of protection and a funding shortfall of $1.2 billion over the next decade.[46]

The land trust and environmental protection movements come together in the most prominent private land conservation organization in the nation, The Nature Conservancy (TNC). Started by biologists concerned about lack of protection for species and habitat, TNC early on forged the basic idea that the best way to protect nature was to buy and manage the land itself, and it has been tremendously successful in this effort. TNC has protected, by direct means such as fee simple purchases, conservation easements, and transfer to public ownership, something on the order of 120 million acres in the United States alone.[47]

Especially over the last two decades, TNC has added broad programs aimed at "community conservation" and advocacy for public conservation of land. The community conservation strategy was based on the logic that TNC holdings could never accomplish the whole job of protecting biodiversity in any landscape, so there was a need to reach out to the surrounding community, by helping ranchers stay in business, working with local communities on open space and habitat protection programs, and promoting land conservation by coalitions of other private and public institutions. TNC has also devoted more attention to local and regional land use planning, becoming in some areas something of a development watchdog group, as a logical approach to protecting its own portfolio of land and the ecological processes that operate across its boundaries.

Improve and Apply Data and Models for Land Use Assessment, Projection, and Decision Making

Although the key ingredients necessary to create a new land use regime in the West are social and institutional, the current proliferation of tools—land use data, digital mapping systems, land use simulation models, and community and land planning and conservation decision support systems—can also improve land use outcomes. Indeed, the growing suite of technical tools has the power not only to improve decision making, but also to further democratize community and land use planning and growth management.

Three categories of analytical planning tools stand out: GIS and similar data mapping tools, land use decision support tools, and community planning tools (although of course these approaches and methods overlap). Perhaps because I am a geographer, I place GIS at the top of the list as, I believe, the most profound innovation.

GIS: More than the Lay of the Land

Geographic information systems (GIS) are revolutionary. They allow people to look at geographic patterns in new ways, to visualize how multiple factors interact on the landscape, and to see the spatial world in which they live from an overhead perspective. Of course, maps have always done this, and they have long been a vital part of land use analysis and planning. Paper-and-ink cartography was limited in several ways, however: hardcopy maps were costly to produce and difficult to amend, and despite decades of effort to create methods whereby hardcopy maps could show layers (planners traditionally created sets of transparent plastic overlays that could be physically lifted off and placed back on a paper base map, thus the term "overlay zone"), the technology was clumsy.

The key innovation of GIS layers and maps is not only their digital, electronic form, so malleable and so quickly redrafted, projected, and distributed, but also the computational and analytical power behind them. A decade ago, if I asked a student to draw a map showing how land ownership, wildlife habitat, and, say, roads or houses were positioned on the landscape, the resulting illustration was both difficult to produce (as line widths, shapes, and colors had to be chosen to allow one layer to show through the other) and difficult to read (and, if it didn't turn out quite right, time-consuming to redraft). Today, with the same data, the map can be created once with help from digital wizards and then iterated in any number of ways; layers can be turned on and off according to the viewer's needs. Landowners or ecologists wanting to know, say, how much and what parts of a parcel contain different habitat types, or lie certain distances from roads, can calculate those values quickly and then map out areas with certain parameters, such as highly valued habitat that is especially far from roads. For a geographer eager to parse out the complexities that make up a landscape, it doesn't get any better than that!

GIS also provides the vessel into which we can finally fit all the land data that simply didn't mesh well and made little geographic sense to users (e.g., parcel legal descriptions and lot numbers). Of course, this takes effort, staff, and machines. We are a long way from full, coordinated, and compatible land use GIS databases for the West; most areas still lack digital parcel or land use maps, and many smaller cities and counties don't have the machines or staff to develop GIS data systems.

Every community in the West should have richly populated, comprehensive GIS databases of land use, plans, population, and other factors, as well as access to tools that allow planning advocates, citizens, and professionals to manipulate, test, extrapolate, and evaluate development patterns at various scales, from local to regional. We're very close to having the suite of modeling tools to do this, but the data are often lacking, and the resources are not always available to apply the tools where they're most needed. The obvious source of land use data, local government, is often problematic. Many local jurisdictions do not even maintain an up-to-date land use map, and even fewer manage to keep parcel maps up to date. Many jurisdictions are overly protective of their land use and parcel data. I have been told by more than one local official that land ownership and parcel data are not public (they patently are public), and many local governments charge a significant price for GIS layers, especially for parcel data, although they were created with public funds.

Although charitable foundations and other NGOs have put some effort into tool development, and even into making GIS systems and data available to environmental and social advocacy groups across the West, there are still too few data and too little effort on land use per se. Progress is being made on ecological and species mapping, but less on land ownership and development data. For most of the West, we simply do not know what lands have been developed, and at what intensity, at a resolution and scale useful to land use analysis and planning. Mapped data on water resources and other infrastructural themes are even more difficult to acquire, yet they are important to anyone wishing to assess the sustainability of development patterns. A concerted effort to put more land use data at the public's disposal is in order.

Finally, GIS can be used to project development and land use into the future and to ask "what if" questions; in this incarnation, GIS blends into the world of planning models. This capability in itself may be the single most important use of GIS in community and regional planning, especially in a rapidly growing region where the future will look very different from the present. It was not long after GIS became commonly available and land use data were digitized that many groups (including my own "Western Futures" research team) began to make maps of the future. Envision Utah mapped out the Wasatch Front urban footprint for 2050 and put its maps on the Web as a "movie," like a weather satellite loop, that began in 1997 and stepped to 2050 as the purple splotch of development grew, then recycled back to 1997.[48] The many compelling future development maps now available include, for example, the Sacramento projections for 2050 (plate 9) and Puget Sound land use forecasts, rather courageously extrapolated to 2100![49]

With data and analyses in hand, land use advocates could match or exceed the analytical capacity of local governments, taking the time that local authorities never seem to have to measure community indicators such as trails and open space, the location of affordable housing and transit, and buildout patterns. Moreover, we have the land use and community development models needed to test alternative development patterns against such measurements and to project community patterns into the future, a future being decided now in every community across the rapidly growing American West.

Specialized Land Use Decision Support Systems

In between GIS applications and full-blown community planning models (described below) range a set of GIS-based systems designed for specific analysis and management goals. The most common of these goals is land conservation.

It was only logical that a tool that could easily compile and analyze multiple geographic features would aid land conservation, which is driven by well-established spatial criteria of land and habitat qualities. Conservation goals almost always include the following criteria: conserve species of interest and their habitats; conserve the largest contiguous areas possible; and conserve lands within a regional strategy that seeks a portfolio of representative habitats, key landscapes, and species. Finally, because land conservation necessarily proceeds over long time frames as finances are arranged and land becomes available, analysts use some measure of "threat," often linked to population growth and the spread of development (mapped as described above), to set priorities.

The Nature Conservancy uses this approach, which it calls "ecoregional planning" or "conservation by design." The ecoregional aspect is founded on TNC's grand plan to build a portfolio of protected lands that represent all the nation's major ecosystems. Its "Conservation by Design" approach includes extensive mapping and analysis of land status, condition, species, and so forth (box 9.4). It took some time for an NGO to grow large enough and accumulate sufficient funding to shift to such a comprehensive approach. TNC's goal is impressive: it is, essentially, the maintenance of the globe's biodiversity through strategic land conservation to encompass whole ecosystems. Its use of GIS and landscape planning and management tools represents a maturation of environmental advocacy that will forever change land use planning in the United States.

Box 9.4

"Conservation by Design": How The Nature Conservancy Protects Whole Landscapes

Conservation organizations have made great strides in developing scientific and strategic approaches to land and habitat conservation, especially for private lands. In a sense, they are more free than public agencies to think and plan across jurisdictional boundaries and to let the needs of species drive their land protection plans. The Nature Conservancy is the leader in strategic land protection planning, applying a system called "Conservation by Design." TNC divides the world into ecoregions and assesses the status of each. The assessment yields a list of protection priorities based on landscape units collected into a portfolio, as well as approaches and outcomes. It depends on the following basic concepts:

- An *ecoregion* is a large unit of land and water typically defined by climate, geology, topography, and associations of plants and animals. Ecoregions, not political boundaries, provide a framework for capturing ecological and genetic variation in biodiversity across a full range of environmental gradients.
- An *ecoregional portfolio*, the end product of ecoregional planning, is a selected set of places that represents the full distribution and diversity of native species, natural communities, and ecosystems in an ecoregion. If managed appropriately, a portfolio will ensure the long-term survival of all native life and natural communities, not only threatened species and communities.
- *Functional conservation areas* conserve focal species, natural communities, ecosystems, and the ecological processes necessary to sustain them over the long term. Functional conservation areas range along a continuum of complexity and scale, from landscapes that seek to conserve a large number of conservation targets at multiple spatial scales to sites that seek to conserve a small number of conservation targets. To conserve wide-ranging and migratory species, functional conservation areas within and across portfolios should be designed as integrated networks.
- *Functional landscapes* represent particularly effective and efficient geographic units for conserving biodiversity within ecoregions. Large, complex, and relatively intact at multiple scales, functional landscapes

continued on next page

Box 9.4 continued

provide an ecological stage on which biodiversity can respond to human or natural disturbances.

TNC sets priorities in two ways: through global major habitat type assessments and through portfolios of conservation areas within and across ecoregions. Ecoregional portfolios represent the full distribution and diversity of native species, natural communities, and ecosystems. Designing ecoregional portfolios is a complex, iterative process based on five steps:

1. Identifying the species, communities, and ecosystems in an ecoregion
2. Setting specific goals for the number and distribution of those conservation targets to be captured in the portfolio
3. Assembling information and relevant data on the location and quality of those conservation targets
4. Designing a network of conservation areas that most effectively meets the conservation goals
5. Identifying high-priority conservation areas, wide-ranging targets, and pervasive threats for conservation action

To address ecoregional priorities and meet ecoregional goals, TNC develops and implements conservation strategies using a "5-S Framework for Conservation Project Management" focused on the following components:

- *Systems*: The focal conservation targets and their key ecological attributes
- *Stresses*: The most serious types of destruction or degradation affecting the conservation targets or their key ecological attributes
- *Sources of stress*: The causes or agents of destruction or degradation
- *Strategies*: The full array of actions necessary to abate the threats or enhance the viability of the conservation targets
- *Success measures*: The monitoring process for assessing progress in abating threats and improving the biodiversity and health of a conservation area

Every ecoregional assessment also identifies priorities that affect multiple conservation areas. These priorities include wide-ranging target species, pervasive critical threats, and institutions and mechanisms that have effects on multiple conservation areas within a given portfolio, among several portfolios, or across geopolitical boundaries. Single-area strategies are typically insufficient to address such multi-area priorities, so TNC designs conservation

Box 9.4 continued

strategies of sufficient scope and scale to address these multi-area priorities. TNC's conservation actions span the spectrum from fee-simple acquisition of land and waters to environmental education; from public policy to joint land and water management agreements. Finally, the process includes long-term monitoring and measures of success.

Source: "Conservation by Design: A Framework for Mission Success" (Arlington, VA: The Nature Conservancy, 2004). Available at: http://www.nature.org/aboutus/howwework/files/cbd_en.

Community Planning Models

As any participant in a planning meeting will attest, planners are consummate modelers, and they love simulations that help them, and citizens, visualize development. In constant need of good ways to project multiple futures to stakeholders, they have adopted everything from physical models, in which blocks of wood that stand in for buildings can be moved around by participants at public workshops, to community mapping, in which focus groups use colored markers and huge sheets of paper to try out ideas for how their community will evolve. Often planners themselves have created the images and maps; indeed, it was common not long ago for most planners to have some skills in design drawing, producing appealing watercolor paintings of everything from landscapes to buildings to streets. Each drawing was, in effect, a simulation. In addition to a certain tendency to artistically glorify the scenery, such approaches (still in use) are subject to the same limits as hand-drawn maps: they are labor-intensive and difficult to produce in iterations that reflect projected changes.

The digital revolution came to the rescue here, too. First, many planners learned to manipulate photographs so that actual images became alternative scenarios (e.g., with more or fewer homes on a hill slope). Then university, agency, and private sector developers began creating computer simulation tools that allowed planners to present complex visual images of alternative designs. Streetscapes could be quickly reformulated with alternative design standards, as everything from the look of buildings to street width to streetside vegetation could be manipulated and re-presented on the fly at public meetings, in focus groups, or even in one-on-one sessions with developers, planning commissioners, and elected officials. Through the 1990s, such models were elaborated so that quantities such as cost, floor area to lot size ratios, and traffic and parking could be linked to designs and could thus become part of the alternative scenarios to be digested by community leaders and the public. The most elaborate models include design, land use, and associated factors such as population, road mileage, and density, as well as environmental effects such as air pollution and urban runoff. One citywide simulation model, SimCity, became a wildly popular computer game.[50]

Several community planning models have been successfully applied in the West, at various scales and with various goals. PLACE^{3}S, a set of planning principles and simulation tools developed jointly by the state energy offices of California, Oregon, and Washington,[51] was used by the Sacramento Area Council of Governments to develop the Sacramento Region Blueprint. PLACE^{3}S employs energy consumption and emissions as a main rubric to assess the suitability (and sustainability) of community design, transportation, and land use patterns, but its GIS-based routines can be extended to assess any significant land use factor; indeed, in the Sacramento case, planners incorporated smart growth principles such as compact development and open space into the modeling.

The key value of PLACE^{3}S is in its application to a public process and to quick assessment of alternatives and their outcomes. The Sacramento Region Blueprint process began, analytically, with a "business-as-usual" scenario that projected the region's land use, transportation, and population out to 2050. It then allowed users to change planning standards and assess the effects of those changes. As the SACOG planners put it, they could "test-drive" smart growth principles. The process was demanding, including dozens of workshops with PLACE^{3}S running on laptop computers ready to test the land use, economic, and social outcomes of alternative development guidelines. The process also included nested scales—that is, neighborhood and regional plans—and their interactions. At the neighborhood scale, which was the basic building block, the process included thirty neighborhood workshops focusing on sixty study areas (attracting over

1,500 participants) and running dozens of scenarios for outcomes such as vehicle miles traveled, pollution, jobs and dwelling units, density, and economic feasibility. Next came county-level workshops for the five counties in the region (Sacramento, Yolo, Placer, Yuba, and Sutter), working with the "business-as-usual" scenario and three alternatives derived from the neighborhood sessions. The focus here was on overall amount of growth, balance of land uses, densities, locations, and resource protection. Finally, alternative scenarios for the entire region were taken through similar workshops and tested against principles and outcomes, culminating in an "electronic town hall" and adoption of a new regional plan.

At the regional scale, the Sacramento Region Blueprint planning process was a pathfinding process worthy of study. SACOG (the region's MPO) was able to go far beyond transportation, making a credible stab at assessing goals and outcomes, for example, for farmland and habitat protection. Its scenarios are based on a future development map. The development footprint can then be used to assess other effects. For example, an important ecological zone in this area is the hardwood oak forest that skirts the Sierra Nevada foothills. The amount of this eco-zone affected by different development footprints can be quickly calculated and displayed.

Another model, CommunityViz, was developed by the Orton Family Foundation specifically to improve planning and decision making in small towns in New England and the Rocky Mountain West grappling with rapid growth and change. As its creators stated, the model's functionalities are designed to achieve community goals:

> CommunityViz provides GIS-based analysis and real-world 3D modeling that allow people to envision land use alternatives and understand their potential impacts, explore options and share possibilities, examine scenarios from all angles—environmental, economic, and social—and feel confident in their decisions.[52]

Orton's goal for CommunityViz was to make it more than planning software. It is "a way of planning that allows people to think and act like citizens—beyond their own backyards and bottom lines—by employing the shared language of visualization." It helps people with different viewpoints and backgrounds engage in collaborative, informed, and equitable decision making about their common future, turning civic innovation into everyday practice.

The usefulness of such models depends on their integrated functionality. CommunityViz and similar software packages can walk the user through, for example, a "buildout analysis." When applied more broadly—to a comprehensive plan, for example (plate 10)—models like CommunityViz allow stakeholders to assign values to certain landscape outcomes, set thresholds, test them against alternative plans, and see the results in various metrics, such as suitability (e.g., for development) or cost (e.g., for roads).

Use of planning models can be demanding and problematic. Subroutines for factors such as traffic or for cost-benefit analysis are only as good as the data and calculations on which they are based. Furthermore, most models operate in a "black box" mode, in which users don't know the underlying rubrics and assumptions that yield certain results. Many of the most controversial questions surrounding community development, such as whether certain types and patterns of development cover their added cost of services or whether some infrastructure investments encourage sprawl, are sensitive to the assumptions necessary to make the models work.

The use of community planning models in actual public workshops also requires a matching of machine power and technical assistance, multiple workstations interacting in real time with the data layers, and compelling display capabilities. This is a tall technical order.

Finally, planners struggle with the challenge of linking the most specialized tools, such as land and habitat conservation models, with broader integrated community planning models. The very best transportation, wildlife habitat, and fiscal models are not readily integrated into comprehensive community planning models; they demand too much data and computational power. The professional planner remains central to drawing together all the technical information needed to make better land use decisions.

Apparatus for a New West

From neighborhood advocacy groups to West-wide smart growth and planning organizations, the institutional capacity for achieving healthier land use planning and development patterns in the West is growing rapidly. Improved local comprehensive plans, community and regional development plans, affordable housing programs, local and regional land conservation efforts, and large-scale ecoregional schemes provide the building blocks for a new development trajectory that conserves resources, wildlife habitat, and open space; promotes more equitable land use and development patterns; and enhances sense of community across the West's dramatic landscapes.

But tools are not sufficient. We also need a new template for land use planning in the West. It has to transcend jurisdictional boundaries and integrate ecological and social well-being. Additionally, a larger social engagement in land use planning is needed, one in which planning is demystified, fully democratized, and taken on as a civic responsibility. Historian Richard White half-joked that planning was boring, but that it was also so important that every citizen should be engaged in it.[53] Of course, planning is not boring at all when it affects your own backyard, and every planner will attest to the fact that any plan, even a minor one, will evoke strong responses from some citizens. The energy is there, the care is there. That energy and care must be channeled into what William Shutkin calls "civic environmentalism," built on "personal relationships and networks, neighborhoods and livelihoods, across geographic, cultural, and political borders" that "links urban, suburban and rural constituencies in the pursuit of shared goals and visions, and enforces the notion that our fates are bound together by place and time."[54]

So, more than ever in its past, the West is rich in ideas for improved land use that enhances the region's natural and social wealth, organizations that seek better planning, and tools for improving land use patterns. But as a democratic process, land use planning needs the participation of individuals who choose to speak out for and work for their communities. The opportunity for those individuals to influence how development plays out across the West's landscapes is greater than ever before.

Endnotes

Introduction

1. The American West is defined in many different ways; here, the focus is on the eleven western states (Arizona, California, Colorado, Idaho, Montana, Nevada, New Mexico, Oregon, Utah, Washington, and Wyoming) and, within them, on the Interior West from the Front Ranges of the Rockies west to the Sierra Nevada and Cascade ranges. Broader definitions sometimes include the Great Plains, which have a markedly different settlement and growth history than the region I explore here, and some analysts include Alaska and Hawaii. I will touch on development issues along the Pacific coast, from San Diego to Seattle, but focus mostly on the interior, where land use battles are shaping the future development regime.

2. This is only a partial list of ski area improvements across the West since the late 1990s. One such expansion, into Vail's Blue Sky Basin, apparently provoked environmental protestors to burn down several on-mountain facilities; see Daniel Glick, *Powder Burn: Arson, Money and Mystery on Vail Mountain* (New York: PublicAffairs, 2001).

3. The *New York Times* Travel section offered a roster of western ski resort growth; see Meg Lukens Noonan, "It Takes a (Bigger, Better) Village," *New York Times*, November 14, 1999.

4. As recently as 1981, journalist Joel Garreau labeled much of the Interior West the "Empty Quarter" (although just ten years later, he wrote about the remarkable growth of "Edge Cities" associated with Denver and Phoenix); see Joel Garreau, *The Nine Nations of North America* (Boston: Houghton Mifflin, 1981); and Joel Garreau, *Edge City: Life on the New Frontier* (New York: Doubleday, 1991).

5. Carl Abbot, *The Metropolitan Frontier* (Tucson: University of Arizona Press, 1993).

6. Charles Wilkinson, "Paradise Revised," in *Atlas of the New West: Portrait of a Changing Region*, ed. William E. Riebsame (New York: W.W. Norton, 1997), 17.

7. Mike Davis, "Las Vegas vs. Nature," in *Reopening the American West*, ed. H. K. Rothman (Tucson: University of Arizona Press, 1998): 59.

8. Andrew J. Hansen, Ray Rasker, Bruce Maxwell, Jay J. Rotella, Jerry

D. Johnson, Andrea Wright Parmenter, Ute Langner, Warren B. Cohen, Rick L. Lawrence, and Matthew P. V. Kraska, "Ecological Causes and Consequences of Demographic Change in the New West," *BioScience* 52 (2002): 151–162.

9. Mathis Wackernagel and William Rees, *Our Ecological Footprint: Reducing Human Impact on the Earth* (Gabriola Island, BC: New Society Publishers, 1996): 7–16.

10. Rutherford H. Platt, *Land Use and Society: Geography, Law, and Public Policy* (Washington, DC: Island Press, 1996).

11. Patricia N. Limerick, "Progress or Decline: Judging the History of Western Expansion," in *A Society to Match the Scenery: Personal Visions of the Future of the American West*, ed. Gary Holthaus, P. N. Limerick, C. F. Wilinson, and E. S. Munson (Niwot: University Press of Colorado, 1991), 44.

Chapter 1

1. Book-length attempts to capture the West's recent demographic, economic, and cultural florescence include William Riebsame, ed., *Atlas of the New West: Portrait of a Changing Region* (New York: W. W. Norton, 1997); and Thomas Michael Power, *Lost Landscapes and Failed Economies: The Search for a Value of Place* (Washington, DC: Island Press, 1996). Regional assessments focused on job growth and rural development in the West appeared in a special issue of *Rural Development Perspectives* 14, no. 2 (1999).

2. *Time* magazine set the tone with a cover story describing the "booming good time" enjoyed by the Rocky Mountain states (Jordan Bonfante, "Boom Time in the Rockies," *Time*, September 6, 1993, 20–27). Journalist Robert D. Kaplan assayed the region's booming economy and technological florescence in *Atlantic Monthly* 282, no. 2 (1998): 37–61. The *New York Times* offered a running commentary throughout the 1990s on the social and ecological effects of the boom in the West, alternating from gloomy to cheery (e.g., Jon Christensen, "In the Sierras, Growth and Preservation Are Not at Odds," *New York Times*, November 30, 1997).

3. *High Country News* publisher Ed Marston most clearly articulated this sentiment in a column entitled "This Boom Will End Like the Others—in a Deep, Deep Bust," *High Country News*, September 5, 1994. See also Patricia N. Limerick, William R. Travis, and Tamar Scoggin, *Boom and Bust in the American West,* Report from the Center no. 4 (Boulder: Center of the American West, University of Colorado at Boulder, 2002).

4. Howard R. Lamar, ed., *The New Encyclopedia of the American West* (New Haven, CT: Yale University Press, 1998), 154; Rodman W. Paul, *Mining Frontiers of the Far West, 1848–1880* (New York: Holt, Rinehart and Winston, 1963).

5. Randall Rhoe, "Environment and Mining in the Mountainous West," in *The Mountainous West: Explorations in Historical Geography*, ed. William Wyckoff and Larry M. Dilsaver (Lincoln: University of Nebraska Press, 1995): 169–193.

6. Michael Williams, "The Last Lumber Frontier?" in Rhoe, *The Mountainous West*, 224–250; Richard White, *It's Your Misfortune and None of My Own: A New History of the American West* (Norman: University of Oklahoma Press, 1991).

7. Debra L. Donahue, *The Western Range Revisited: Removing Livestock from Public Lands to Conserve Native Biodiversity* (Norman: University of Oklahoma Press, 1999).

8. William G. Robbins, *Colony and Empire: The Capitalist Transformation of the American West* (Lawrence: University of Kansas Press, 1994), 62.

9. John Opie, *The Law of the Land: Two Hundred Years of American Farmland Policy* (Lincoln: University of Nebraska Press, 1987).

10. Patricia N. Limerick, *The Legacy of Conquest: The Unbroken Past of the American West* (New York: W. W. Norton, 1987), 55.

11. Opie, *Law of the Land*.

12. Metropolitan Denver, along with the entire Colorado Front Range urban area, sits in what Major Stephen H. Long called the Great American Desert after he visited the area in 1820. See White, *It's Your Misfortune*, 121.

13. Charles Wilkinson, *Crossing the Next Meridian: Land, Water, and the Future of the West* (Washington, DC: Island Press, 1992).

14. White, *It's Your Misfortune*, 473–477, 483–487.

15. White, *It's Your Misfortune*, 497.

16. Limerick, *Legacy of Conquest*, 136–139.

17. Limerick et al., *Boom and Bust*.

18. Charles Wilkinson, *Fire on the Plateau: Conflict and Endurance in the American Southwest* (Washington, DC: Island Press, 1999).

19. Patricia N. Limerick, Claudia Puska, Andrew Hildner, and Eric Skovsted, *What Every Westerner Should Know about Energy* Report from the Center no. 4 (Boulder: Center of the American West, University of Colorado at Boulder, 2003).

20. Wilkinson, *Fire on the Plateau*.

21. Andrew Gulliford, *Boomtown Blues: Colorado Oil Shale, 1885–1985* (Niwot: University Press of Colorado, 1989).

22. Limerick et al., *What Every Westerner Should Know about Energy*.

23. Limerick et al., *What Every Westerner Should Know about Energy*.

24. Riebsame, *Atlas of the New West*, 108.

25. The region's growing attractiveness to domestic migrants is well documented for both urban and rural areas. See, for example, Edward L.

Glaeser and Jesse M. Shapiro, "City Growth and the 2000 Census: Which Places Grew, and Why," Survey Series (Washington, DC: Brookings Institution, Center on Urban and Metropolitan Policy, 2001); George Masnick, "America's Shifting Population: Understanding Migration Patterns across the West," *The Rocky Mountain West's Changing Landscape* 2, no. 2 (Winter/Spring 2001): 8–14; John B. Cromartie and John M. Wardwell, "Migrants Settling Far and Wide in the Rural West," *Rural Development Perspectives* 14, no. 2 (1999), 2–8; Gundars Rudzitis, "Amenities Increasingly Draw People to the Rural West," *Rural Development Perspectives* 14, no. 2 (1999): 9–13.

26. Power, *Lost Landscapes*, 5.

27. Thomas Michael Power and Richard N. Barrett, *Post-Cowboy Economics: Pay and Prosperity in the New American West* (Washington, DC: Island Press, 2001), 147.

28. In a 2003 case, the Colorado Water Conservation Board argued that Golden's claim on certain flows of Clear Creek for use in a whitewater park violated water law because that application did not constitute "beneficial use" of water, traditionally taken to mean only water diverted for crops, industry, or municipal applications. Golden's water attorney, Glen Porzak, won the case in the Colorado Supreme Court by showing that the whitewater park brought $23 million to Golden's economy annually, more than the water yielded in most other uses. See American Whitewater, "Judge Orders Water for Golden's Whitewater Park in CO," http://www.americanwhitewater.org/archive/article/202/ (accessed March 1, 2004).

29. Power, *Lost Landscapes*.

30. Bonfante, "Boom Time in the Rockies."

31. Bonfante, "Boom Time in the Rockies," 23.

32. "Pacific Northwest Paradises," *Newsweek*, May 20, 1996, 54.

33. Raye C. Ringholz, *Paradise Paved: The Challenge of Growth in the New West* (Salt Lake City: University of Utah Press, 1996); Hal K. Rothman, *Devil's Bargains: Tourism in the Twentieth-Century American West* (Lawrence: University Press of Kansas, 1998).

34. George Wuerthner, "Subdivisions versus Agriculture," *Conservation Biology* 8, no. 3 (1994): 905–908.

35. William B. Beyers, "Employment Growth in the Rural West from 1985 to 1995 Outpaced the Nation," *Rural Development Perspectives* 14, no. 2 (1999): 38–43.

36. Ray Rasker, "Your Next Job Will Be in Services. Should You Be Worried?" in *Across the Great Divide: Explorations in Collaborative Conservation and the American West*, ed. Phillip Brick, Donald Snow, and Sarah Van de Wetering (Washington, DC: Island Press, 2001), 51–57.

37. Power, *Lost Landscapes*, 237.

38. Power and Barrett, *Post-Cowboy Economics*, 147.

39. Power and Barrett, *Post-Cowboy Economics*, 147.

40. Power and Barrett, *Post-Cowboy Economics*, 147.

41. Sierra Business Council, *Planning for Prosperity: Building Successful Communities in the Sierra Nevada* (Truckee, CA: Sierra Business Council, 1997); Patrick C. Jobes, *Moving Nearer to Heaven: The Illusions and Disillusions of Migrants to Scenic Rural Places* (Westport, CT: Praeger Publishers, 2000).

42. Limerick et al., *Boom and Bust*; Power and Barrett, *Post-Cowboy Economics*. The sharp decline of the high-tech economy in 2002–2003 cast some doubt on this outlook. But a series of studies by the Milken Institute, an economic think tank in California, argue that high-tech and economic diversification have indeed stabilized the economy of states such as Colorado, Utah, California, and Arizona. See, for example, Ross DeVol and Perry Wong, "America's High-Tech Economy: Growth, Development and Risks for Metropolitan Areas" (Santa Monica, CA: Milken Institute, 1999). The Milken Institute held firm in this argument even as the national recession struck; see Ross DeVol, with Rob Koepp and Frank Fogelbach, "State Technology and Science Index: Comparing and Contrasting California" (Santa Monica, CA: Milken Institute, 2002). The Milken Institute's reports are available on its Web site at http://www.milkeninstitute.org/publications/publications.taf (accessed March 3, 2004).

43. The National Bureau of Economic Research maintains a "recession timing committee," which offers a nonpartisan definition of economic cycles; see Business Cycle Dating Committee, National Bureau of Economic Research, "The NBER's Recession Dating Procedure," http://www.nber.org/cycles/recessions.html (accessed March 3, 2004).

44. Trent Siebert, "Growth Debate Cited in Slowdown," *Denver Post*, September 24, 2001. Amazingly, the Colorado legislature was in special session and seriously considering several growth management bills just four months earlier, in May 2001.

45. Siebert, "Growth Debate Cited in Slowdown."

46. Rothman, *Devil's Bargains*.

47. Several studies show this problem in rural areas; see, for example, the Sonoran Institute's studies for Custer County, CO, available at http://www.sonoran.org/programs/custercountypilgrimage.html (accessed April 9, 2006).

48. Michael F. Logan, *Fighting Sprawl and City Hall: Resistance to Urban Growth in the Southwest* (Tucson: University of Arizona Press, 1995).

49. The CEO of Level 3, a telecommunications firm that built its head-

quarters between Denver and Boulder, chose the location because the 5,000 employees he hoped to lure wanted to live near skiing and fishing while still having the benefits of a medium-sized city at their doorstep, as described in chapter 2.

50. Peter D. Nichols, Megan K. Murphy, and Douglas S. Kenney, *Water and Growth in Colorado: A Review of Legal and Policy Issues* (Boulder: Natural Resources Law Center, University of Colorado School of Law, 2001).

51. This was the surprising conclusion of a study of the potential effects of a severe, sustained drought in the basin of the Colorado River, arguably the most stressed river in the West. See William B. Lord, James F. Booker, David M. Getches, Benjamin L. Harding, Douglas S. Kenney, and Robert A. Young, "Managing the Colorado River in a Severe, Sustained Drought: An Evaluation of Institutional Options," *Water Resources Bulletin* 31, no. 5 (1995): 939–944.

52. Gundars Rudzitis, *Wilderness and the Changing American West* (New York: John Wiley and Sons, 1996).

53. Timothy P. Duane, *Shaping the Sierra: Nature, Culture, and Conflict in the Changing West* (Berkeley: University of California Press, 1998).

Chapter 2

1. The micropolitan category and other new elements in the Office of Management and Budget and Census Bureau's delineation of metropolitan areas are nicely described in a Brookings Institution "field guide": William H. Frey, Jill H. Wilson, Alan Berube, and Audrey Singer, "Tracking Metropolitan America into the 21st Century: A Field Guide to the New Metropolitan and Micropolitan Definitions," Living Census Series (Washington, DC: Brookings Institution, 2004), available at http://www.brookings.edu/metro/pubs/20041115_metrodefinitions.htm (accessed March 10, 2006).

2. Richard White, *It's Your Misfortune and None of My Own: A New History of the American West* (Norman: University of Oklahoma Press, 1991); Carl Abbott, *The Metropolitan Frontier: Cities in the Modern American West* (Tucson: University of Arizona Press, 1993).

3. David Rusk, *Cities Without Suburbs* (Washington, DC: Woodrow Wilson Center Press, 1993).

4. Harvey Molotch, "The City as a Growth Machine: Toward a Political Economy of Place," *American Journal of Sociology* 82 (1976): 309–332.

5. Hal K. Rothman, *Devil's Bargains: Tourism in the Twentieth-Century American West* (Lawrence: University Press of Kansas, 1998), 168–286.

6. Robert E. Land, "Office Sprawl: The Evolving Geography of Business," Survey Series (Washington, DC: Brookings Institution, Center on

Urban and Metropolitan Policy, 2000). See also Joel S. Hirschhorn, *Growing Pains: Quality of Life in the New Economy* (Washington, DC: National Governors' Association, 2000), 15.

7. John B. Cromartie and John M. Wardwell, "Migrants Settling Far and Wide in the Rural West," *Rural Development Perspectives* 14, no. 2 (1999): 2–8.

8. William B. Beyers, "Employment Growth in the Rural West from 1985 to 1995 Outpaced the Nation," *Rural Development Perspectives* 14, no. 2 (1999): 38–43.

9. Raye C. Ringholz, *Paradise Paved: The Challenge of Growth in the New West* (Salt Lake City: University of Utah Press, 1996); Edward Abbey, *Desert Solitaire: A Season in the Wilderness* (New York: Ballantine Books, 1968), 52.

10. Ringholz, *Paradise Paved*, ix–x.

11. Rothman, *Devil's Bargains*.

12. This well-known phenomenon is best illustrated in the Roaring Fork Valley below Aspen, Colorado, where resort real estate values have trickled down-valley, turning former middle-class towns, such as El Jebel, into resort settings. Chapter 6 diagnoses this pattern in this and other resort areas.

13. Cromartie and Wardwell, "Migrants Settling Far and Wide."

14. Kristopher M. Rengert and Robert E. Lang, "Cowboys and Cappuccino: The Emerging Diversity of the Rural West," Census Note 04 (Washington, DC: Fannie Mae Foundation, 2001).

15. Bradley J. Gentner and John A. Tanaka, "Classifying Federal Public Land Grazing Permittees," *Journal of Range Management* 55 (2002): 2–11.

16. D. P. Smith and D. A. Phillips, "Socio-Cultural Representations of Greentrified Pennine Rurality," *Journal of Rural Studies* 17 (2001): 457–469.

17. Peter B. Nelson, "Rural Restructuring in the American West: Land Use, Family and Class Discourses," *Journal of Rural Studies* 17 (2001): 395–407. See also Peter Decker, *Old Fences, New Neighbors* (Tucson: University of Arizona Press, 1998).

18. See Charles Wilkinson's now classic criticism of outdated public lands policy, *Crossing the Next Meridian: Land, Water, and the Future of the West* (Washington, DC: Island Press, 1992).

19. Andrew J. Hansen, Ray Rasker, Bruce Maxwell, Jay J. Rotella, Jerry D. Johnson, Andrea Wright Parmenter, Ute Langner, Warren B. Cohen, Rick L. Lawrence, and Matthew P. V. Kraska, "Ecological Causes and Consequences of Demographic Change in the New West," *BioScience* 52, no. 2 (2002): 151–162.

20. Cromartie and Wardwell, "Migrants Settling Far and Wide."

21. Hal K. Rothman, ed., *Re-opening the American West* (Tucson: University of Arizona Press, 1998).

22. Development proponents in Denver, Boise, Phoenix, and essentially every other western city point to mountain views, plus nearby amenities such as national parks and forests, trail systems, and ski resorts, as reasons to locate a business in those cities.

23. William B. Beyers and Peter B. Nelson, "Contemporary Development Forces in the Nonmetropolitan West: New Insights from Rapidly Growing Communities," *Journal of Rural Studies* 16 (2000): 459–474.

24. George Masnick, "America's Shifting Population: Understanding Migration Patterns across the West," *The Rocky Mountain West's Changing Landscape* 2, no. 2 (Winter/Spring 2001): 8–14.

25. The literature on immigration is extensive. A careful look at its impacts in California is provided by geographer William A. V. Clark in *The California Cauldron: Immigration and the Fortunes of Local Communities* (New York: Guilford Press, 1998). See also William A. V. Clark, "Immigration and California Communities," *Backgrounder* (Los Angeles: Center for Immigration Studies, 1999), at http://www.cis.org/articles/1999/back299.html (accessed May 1, 2002). The Centers for Disease Control and Prevention's National Vital Statistics System reports that international immigration appears to have increased U.S. fertility rates; see Joyce A. Martin, Brady E. Hamilton, Paul D. Sutton, Stephanie J. Ventura, Fay Menacker, and Martha L. Munson, "Births: Final Data for 2003," National Vital Statistics Report 54, no. 2 (Hyattsville, MD: National Center for Health Statistics, 2005), http://www.cdc.gov/nchs/data/nvsr/nvsr54/nvsr54_02.pdf (accessed April 6, 2006).

26. Fertility data are collected by the Centers for Disease Control and Prevention and are published in their National Vital Statistics System (NVSS) reports. These data are from Martin et al., "Births: Final Data for 2003."

27. David Savageau, *Retirement Places Rated*, 4th ed. (New York: Macmillan Travel, 1995).

28. William H. Frey and Ross C. DeVol, "America's Demography in the New Century: Aging Baby Boomers and New Immigrants as Major Players," Policy Brief (Santa Monica, CA: Milken Institute, 2000), available at http://www.milkeninstitute.org/publications/publications.taf?function=detail&ID=54&cat=PBriefs (accessed November 3, 2006).

29. Brandon Loomis, "Sun, Fun, and Sprawl," *Salt Lake Tribune*, March 4, 2001. Loomis described the retirement boom around Las Vegas as an example of why the Interior West states are growing so fast; he also coined the phrase "manifest-destiny golfers."

30. Rengert and Lang, "Cowboys and Cappuccino."

31. The projections are available from the U.S. Census Bureau at

http://www.census.gov/population/www/projections/popproj.html (accessed April 8, 2006).

32. Quoted in Aldo Svaldi, "Job Falloff Impacting State Growth," *Denver Post*, March 10, 2002.

33. John S. Sanko, "Californians, Texans Top Newcomers," *Rocky Mountain News*, June 5, 2002.

34. Quoted in Robert Sanchez, "Population Growth Slogs Ahead Despite State's Job Losses," *Denver Post*, April 20, 2006.

35. After the publication of his book *Lost Landscapes and Failed Economies: The Search for a Value of Place* (1996), Thomas Michael Power spoke to community groups across the West, warning them that adherence to the old economic model—which kept them chasing the "smokestacks" of failing Old West industries—was not only futile, but also lessened their chances of capturing the best elements of the new economy. I observed the skepticism with which residents responded to him, based on a sort of economic fundamentalism that holds that a community must export raw resources or manufactured items to survive. Many residents told him that the "New Economy" of services was worse than the old. Power and others have only slowly chipped away at these beliefs even as the detritus of the old economy accumulates on the western landscape.

36. Several economists and demographers have taken a stab at disentangling the jobs-people connection. The New West School recognizes that people need employment, but asserts that quality of life has more power to attract people—who then bring jobs and income with them—than the Old West or "base economy" school—which simply assumes that jobs come first—is willing to admit. See, for example, Alexander C. Vias, "Jobs Follow People in the Rural Rocky Mountain West," *Rural Development Perspectives* 14, no. 2 (1999): 49–61; Ray Rasker, "Your Next Job Will Be in Services. Should You Be Worried?" in *Across the Great Divide: Explorations in Collaborative Conservation and the American West*, ed. Phillip Brick, Donald Snow, and Sarah Van de Wetering (Washington, DC: Island Press, 2001), 51–58; and Thomas Michael Power and Richard N. Barrett, *Post-Cowboy Economics: Pay and Prosperity in the New American West* (Washington, DC: Island Press, 2001).

37. A growing literature supports this argument, including Peter B. Nelson and William B. Beyers, "Using Economic Base Models to Explain New Trends in Rural Income," *Growth and Change* 29 (1998): 321–344; and Vias, "Jobs Follow People in the Rural Rocky Mountain West."

38. Jim Howe, Ed McMahon, and Luther Propst, *Balancing Nature and Commerce in Gateway Communities* (Washington, DC: Island Press, 1997); David

J. Snepenger, Jerry D. Johnson, and Raymond Rasker, "Travel-Stimulated Entrepreneurial Migration," *Journal of Travel Research* 33 (1995): 51–57.

39. Patrick C. Jobes, *Moving Nearer to Heaven: The Illusions and Disillusions of Migrants to Scenic Rural Places* (Westport, CT: Praeger Publishers, 2000).

40. Patricia Gober, Kevin E. McHugh, and Dennis Leclerc, "Job-Rich but Housing-Poor: The Dilemma of a Western Amenity Town," *Professional Geographer* 45 (1993): 12–20.

41. Snepenger et al., "Travel-Stimulated Entrepreneurial Migration."

42. Ray Rasker, Ben Alexander, Jeff van den Noort, and Rebecca Carter, "Prosperity in the 21st Century West: The Role of Protected Public Lands" (Tucson: Sonoran Institute, 2004), available at http://www.sonoran.org/programs/prosperity.html (accessed November 2, 2006).

43. Howe et al., *Balancing Nature and Commerce*.

44. Judith Kohler, "Energy Boom Jarring West's Outdoors-Dependent Economies," *Daily Camera* (Boulder, CO), December 24, 2003.

45. Jennifer Beauprez, "Level 3 Shows Off New Home," *Denver Post*, November 9, 1993.

46. Sandra Fish, "State's Population Tops 4 Million," *Daily Camera*, December 29, 1999.

47. Joint Venture Silicon Valley Network, *The Future of Bay Area Jobs: The Impact of Offshoring and Other Key Trends* (San Jose: Joint Venture Silicon Valley Network, 2004). Joint Venture also conducts an annual assessment of economics and quality of life, tracking factors like housing and open space that affect desirability of the Bay Area among high-tech workers. Their many studies are available on their Web site at http://www.jointventure.org (accessed December 1, 2006).

48. Timothy P. Duane, *Shaping the Sierra: Nature, Culture, and Conflict in the Changing West* (Berkeley: University of California Press, 1998).

49. Glenn V. Fuguitt, David L. Brown, and Calvin L. Beale, *The Population of Rural and Small Town America* (New York: Russell Sage, 1989).

50. Beyers and Nelson, "Contemporary Development Forces"; Rasker et al., "Prosperity in the 21st Century West."

51. Beyers and Nelson, "Contemporary Development Forces."

52. Ann Markusen, *Regions: The Economics and Politics of Territory* (Totowa, NJ: Rowman and Littlefield, 1987).

53. Molotch, "The City as a Growth Machine."

54. Rothman, *Devil's Bargains*.

55. Markusen, *Regions*, 123–125.

56. This "growth machine" is well documented, seminally by John R. Logan and Harvey L. Molotch, *Urban Fortunes: The Political Economy of Place*

(Berkeley: University of California Press, 1987). Michael Logan lays out the growth machine's urban incarnation in the West for Tucson in *Fighting Sprawl and City Hall: Resistance to Urban Growth in the Southwest* (Tucson: University of Arizona Press, 1995), and Rothman details the growth machine in small towns and resorts in *Devil's Bargains*.

57. Jim Greer, "The Courtship of Boeing," *Denver Post*, May 11, 2001.

58. See http://www.stoplegacyhighway.org (accessed December 1, 2006). According to its critics, the highway will destroy wetlands, threaten farmland, worsen air pollution, and induce sprawl; Mark P. Couch, "E-470 Anchors Land Targeted for Next Suburban Migration," *Denver Post*, July 19, 2001.

59. For Duane's longer list, see *Shaping the Sierra*, 48–54.

60. Kenneth T. Jackson, *Crabgrass Frontier: The Suburbanization of the United States* (New York: Oxford University Press, 1985).

61. Arthur C. Nelson, "Characterizing Exurbia," *Journal of Planning Literature* 6 (1992): 352.

62. National Association of Home Builders and National Association of Realtors®, "Joint News Release: Survey Suggests Market-Based Vision of Smart Growth," April 22, 2002, Washington, DC. See http://www.realtor.org/publicaffairsweb.nsf/Pages/SmartGrowthSurvey02?OpenDocument (accessed November 12, 2002).

63. National Association of Home Builders, press release, April 22, 2002.

64. Dowell Myers and Elizabeth Gearin, "Current Preferences and Future Demand for Denser Residential Environments," *Housing Policy Debate* 12, no. 4 (2001): 633–659.

65. Prudential Utah Real Estate ad for Wolf Creek Ranch; details at http://www.wolfcreekranch.com (accessed March 15, 2002). The advertised "forest" refers to the subdivision's private access to the Uinta National Forest.

66. Ad for Daniel's Gate, *Rocky Mountain News*, June 9, 2001; http://www. danielsgate.com (accessed March 15, 2002).

67. The new school of metropolitan studies started with David Rusk's *Cities without Suburbs* (Washington, DC: Woodrow Wilson Center Press, 1993). Myron Orfield's *American Metropolitics: The New Suburban Reality* (Washington, DC: Brookings Institution Press, 2002) was an especially convincing geographic analysis of urban form, showing how wealth shifted to suburbs. See also Andres Duany, Elizabeth Plater-Zyberk, and Jeff Speck, *Suburban Nation: The Rise of Sprawl and the Decline of the American Dream* (New York: North Point Press, 2000); and Bruce Katz, ed., *Reflections on Regionalism* (Washington, DC: Brookings Institution Press, 2000).

68. Peter D. Nichols, Megan K. Murphy, and Douglas S. Kenney most recently made this argument for Colorado in *Water and Growth in Colorado: A Review of Legal and Policy Issues* (Boulder: Natural Resources Law Center, University of Colorado School of Law, 2003).

69. This pattern is described in detail in W. E. Riebsame, H. Gosnell, and D. M. Theobald, "Land Use and Landscape Change in the U.S. Rocky Mountains I: Theory, Scale and Pattern," *Mountain Research and Development* 16 (1996): 395–405.

70. Measure 37 applied "takings" principles to land use restrictions in Oregon, requiring local government to compensate property owners for lost value; see 1000 Friends of Oregon, "Measure 37: Questions and Answers about Oregon's Land Use Planning Program," http://www.friends.org/issues/M37/documents/M37-Impacts-2006-03-06.pdf; and Oregonians in Action, "Measure 37," http://measure37.com/ (accessed April 9, 2006). The measure is a major setback for statewide land use planning in Oregon, and in response, the governor initiated a full-scale review of three decades of state land use policy, referred to as the "Big Look"; see http://www.lcd.state.or.us/LCD/BIGLOOK/index.shtml (accessed December 1, 2006).

71. See Duane, *Shaping the Sierra*, especially chaps. 9 and 10.

72. Duane, *Shaping the Sierra*, 350, 351, 360–363, 367, 372–373, and 388–390.

73. Measure 37 proponents say they will export their property rights movement to other western states.

74. The antiplanning movement is composed of both traditional local development interests and an intellectual school of thought that finds most government land use planning to be inefficient and antidemocratic, arguments made forcefully by Randal O'Toole in *The Vanishing Automobile and Other Urban Myths* (Bandon, OR: Thoreau Institute, 2001) and offered in an antiplanning manifesto by the Lone Mountain Coalition, *The Lone Mountain Compact: Principles for Preserving Freedom and the Livability of America's Cities and Suburbs* (Bozeman, MT: Political Economy Research Center, 2000). Available at http://www.pacificresearch.org/pub/sab/enviro/ lonemtn. html (accessed December 3, 2006). See also Jane S. Shaw and Ronald D. Utt, *A Guide to Smart Growth: Shattering Myths, Providing Solutions* (Bozeman, MT: Political Economy Research Center/Heritage Foundation, 2000).

75. Greater Yellowstone Coalition, "Montana Supreme Court Sides with GYC at Duck Creek," *Greater Yellowstone Newsletter*, Late Summer, 2001.

Chapter 3

1. See, for example, Mathis Wackernagel and William Rees, *Our Ecological Footprint: Reducing Human Impact on the Earth* (Gabriola Island, BC: New Society Press, 1996), 3, 9.

2. The long-term openness of state lands is not necessarily ensured, given policies that allow for sale to private entities. Referenda in Colorado and Arizona recently required their state land agencies to shift state policy more toward preservation than development or sale. Colorado voters required, via a constitutional amendment in 2000, that at least 300,000 acres of the total 3.2 million acres of state lands be retained in public ownership and maintained as conservation lands. Arizona voters passed a bond issue in 2004 to protect some of the 9 million acres of state land, especially near cities.

3. Colorado's governor asked the State Land Board in May 2005 to map essentially every protected acre in the state; a two-year project was funded. The Land Trust Alliance (http://www.lta.org) has rough numbers of protected acres by state, as reported by local land trusts, but few geographic details; indeed, some land trusts are hesitant to fully reveal land protection patterns.

4. Indeed, data are reported for the county in which the owner/operator resides, even if the actual land is in another, often adjacent, county.

5. The first quote is from the Biodiversity Project's "Farmland Loss at a Glance," http://www.biodiversityproject.org/mediakit/Sprawl_1B_farmland_loss.pdf (accessed July 7, 2005). The second quote is from the American Farmland Trust, *Rocky Mountain Agricultural Landowners Guide to Conservation and Sustainability* (Washington, DC: American Farmland Trust, 2006), 1. Essentially all such claims rely on the Census of Agriculture, which does show decreasing farmland in most parts of the United States, but tells us nothing about the uses into which those acres went.

6. Samuel R. Staley, "The 'Vanishing Farmland' Myth and the Smart-Growth Agenda," Policy Brief 12 (Los Angeles: Reason Public Policy Institute, 2000), available at http://www.reason.org/pb12.pdf (accessed July 21, 2006).

7. Explanations of NRI sampling and measurement procedures, as well as definitions of terms such as "built-up," have changed over time, but the publicly available information for the current (2003) NRI is much better than for previous ones, and the NRI's Web resources have been greatly improved; see http://www.nrcs.usda.gov/technical/NRI (accessed July 18, 2005).

8. This is the latest year for which state-level land cover data are available from the NRI. The data are in U.S. Department of Agriculture, "Summary

Report: 1997 National Resources Inventory (revised December 2000)," Washington, DC: Natural Resources Conservation Service / Ames: Statistical Laboratory, Iowa State University, 2000), available at http://www.nrcs.usda.gov/technical/NRI/1997/summary_report/ (accessed July 18, 2005).

9. See USDA, "Summary Report: 1977." The NRCS reported in 2001 that because it was revising previous estimates of urbanized areas downward, too, the trend toward increasing urbanization was still valid. Sprawl skeptics jumped on the NRI revisions, arguing that they showed that the data were biased and unreliable; see, for example, Randal O'Toole's response in "Vanishing Auto Update no. 2: Revised Natural Resources Inventory Is Still Flawed," Thoreau Institute, http://www.ti.org/vaupdate02.html (accessed July 11, 2005).

10. We calculated housing density for all census block groups in the eleven western states and extrapolated development out to 2040 based on population estimates. For the results, see W. R. Travis, D. M. Theobald, G. W. Mixon, and T. W. Dickinson, "Western Futures: A Look into the Patterns of Land Use and Future Development in the American West," Center of the American West Report no. 6 (Boulder: University of Colorado, 2005), available at http://www.centerwest.org/futures.

11. The valley's exurbanization, driven by "disgruntled folks leaving Missoula's air pollution and traffic jams" plus "West Coasters," was lamented in John B. Wright's *Montana Ghost Dance: Essays on Land and Life* (Austin: University of Texas Press, 1998), 131–156.

12. Christopher J. Duerksen and James van Hemert, *True West: Authentic Development Patterns for Small Towns and Rural Areas* (Chicago: Planners Press/American Planning Association, 2003), 186.

13. David M. Theobald, "Placing Exurban Land Use Change in a Human Modification Framework," *Frontiers in Ecology and Environment* 2, no. 3 (2004): 139–144.

14. The Sierra Nevada Ecosystem Project, Executive Summary, http://ceres.ca.gov/snep/pubs/web/default.html (accessed July 16, 2005), 2.

15. Andrew J. Hansen, Ray Rasker, Bruce Maxwell, Jay J. Rotella, Jerry D. Johnson, Andrea W. Parmenter, Ute Langner, Warren B. Cohen, Rick L. Lawrence, and Matthew P. V. Kraska, "Ecological Causes and Consequences of Demographic Change in the New West," *BioScience* 52 (2002): 151–162.

16. Andrew J. Hansen and Jay J. Rotella, "Biophysical Factors, Land Use, and Species Viability in and around Nature Reserves," *Conservation Biology* 16 (2001): 1112–1122.

17. Salmon recovery efforts by the National Marine Fisheries Service (NMFS) (under the Endangered Species Act) began in the 1990s to focus on reducing the effects of urban sprawl, roads, and commercial developments. In response, the National Association of Home Builders (NAHB) joined the cadre of groups fighting salmon protection. NAHB successfully sued the NMFS over their critical habitat designations for nineteen endangered salmon runs in 2002, claiming that the designations unduly restricted private land use.

Chapter 4

1. Carl Abbott, *The Metropolitan Frontier* (Tucson: University of Arizona Press, 1993); William G. Robbins, *Colony and Empire: The Capitalist Transformation of the American West* (Lawrence: University Press of Kansas, 1994).

2. Lawrence H. Larsen, *The Urban West at the End of the Frontier* (Lawrence: Regents Press of Kansas, 1978).

3. Peter Wiley and Robert Gottlieb, *Empires in the Sun: The Rise of the New American West* (Tucson: University of Arizona Press, 1982).

4. David Howard Bain, *Empire Express: Building the First Transcontinental Railroad* (New York: Viking, 1999), 370.

5. Charles N. Glaab, "Visions of Metropolis: William Gilpin and Theories of City Growth in the American West," *Wisconsin Magazine of History* 45 (1961): 21–31.

6. Charles Wilkinson, *Fire on the Plateau: Conflict and Endurance in the American Southwest* (Washington, DC: Island Press, 1999), chap. 9.

7. Hal Rothman and Mike Davis, *The Grit beneath the Glitter: Tales from the Real Las Vegas* (Berkeley: University of California Press, 2002).

8. Quoted in David Clayton, "Las Vegas Goes for Broke," *Planning*, September 1995, 5.

9. Clayton, "Las Vegas Goes for Broke."

10. Definitive studies of this losing battle are collected in Robert W. Burchell, Anthony Downs, Barbara McCann, and Rahan Mukherji, *Sprawl Costs: Economic Impacts of Unchecked Development* (Washington, DC: Island Press, 2005).

11. One could argue that serious talk of secession by larger L.A. suburban cities marks the beginning of the end of this growth scheme: the center cannot hold. But secession fever, which first struck the San Fernando Valley in the 1970s, cooled as the fiscal effects of not being part of the L.A. omnibus were assessed. See Paul Shigley and William Fulton, "L.A. Unbounded," *Planning*, October 2002, 28–30.

12. William Fulton, Rolf Pendall, Mai Nguyen, and Alicia Harrison, *Who*

Sprawls Most? How Growth Patterns Differ across the U.S. (Washington, DC: Brookings Institution, 2001).

13. Peter Calthorpe and William Fulton, *The Regional City: Planning for the End of Sprawl* (Washington, DC: Island Press, 2001), 9–10.

14. Robert E. Lang, "Open Spaces, Bounded Places: Does the American West's Arid Landscape Yield Dense Metropolitan Growth?" *Housing Policy Debate* 13, no. 4 (2003): 755–778.

15. Robert Bruegmann, *Sprawl: A Compact History* (Chicago: University of Chicago Press, 2005), 220.

16. Randal O'Toole, *The Vanishing Automobile and Other Urban Myths: How Smart Growth Will Harm American Cities* (Bandon, OR: Thoreau Institute, 2001): 392–396; see also *Electronic Drummer,* the Web site of the Thoreau Institute, http://www.ti.org, where O'Toole offers extensive updates to *The Vanishing Automobile*.

17. G. Galster, R. Hanson, M. R. Ratcliffe, H. Wolman, S. Coleman, and J. Freihage, "Wrestling Sprawl to the Ground: Defining and Measuring an Elusive Concept," *Housing Policy Debate* 12, no. 4 (2001): 681–717.

18. Yan Song and Gerrit-Jan Knapp, "Measuring Urban Form: Is Portland Winning the War on Sprawl?" *Journal of the American Planning Association* 70 (2004): 210–225.

19. Several conservative think tanks have issued policy papers criticizing urban growth boundaries, smart growth, New Urbanism, and other ideas aimed at curbing sprawl. See, for example, Wendell Cox, "The Anti-Sprawl War on the Suburbs: False Diagnosis, Hopeless Policies," *Veritas—a Quarterly Publication of Public Policy in Texas*, Spring 2000, http://www.texas policy.com/pdf/2000-veritas-1-1-sprawl.pdf (accessed July 17, 2004). Two planning professors who routinely add to the critique of antisprawl arguments are Peter Gordon and Harry W. Richardson, School of Policy, Planning and Development, University of Southern California; see their "Critiquing Sprawl's Critics," *Policy Analysis* no. 365 (January 24, 2000), Cato Institute, available at http://www-rcf.usc.edu/~pgordon/pdf/pa365.pdf (accessed November 2, 2006). See also Randal O'Toole, "Dense Thinkers," *Reason* 30, no. 8 (January 1999): 44–52, http://reason.com/news/show/30875.html (accessed December 3, 2004).

20. Robert Lang, in "Open Spaces, Bounded Places" (757), applauds planner William Fulton's use of the term "dense sprawl" and he also likes urban modeler John Landis's use of the "dense onion" analogy to describe layer after layer of similar-density development added to the edge of Southern California's metro area.

21. Bruegmann, *Sprawl: A Compact History*, 65.

22. Burchell et al., *Sprawl Costs*, 13.

23. Contemporary sprawl, although perhaps even less dense and more inefficient than its predecessor patterns, does seem to be an extension and intensification of trends obvious at least since the proliferation of highways in the post–World War II era, which allowed a mobile middle class to move away from city cores, commuting to work while living in a semipastoral setting—what historian Kenneth Jackson called, with a bit of critical satire, the "crabgrass frontier." See Kenneth T. Jackson, *Crabgrass Frontier: The Suburbanization of the United States* (New York: Oxford University Press, 1985). Jackson, by the way, traced the longing to live just outside the city in a less urban setting to ancient times, citing a clay tablet expressing suburban preferences among residents of Persia in 539 B.C. Taking up the challenge of explaining sprawl, historian Robert Bruegmann, in *Sprawl: A Compact History*, also traces its ancient roots and, like Jackson, argues that it stems in large part from enduring individual preferences.

24. Robert E. Lang, "Office Sprawl: The Evolving Geography of Business," Survey Series (Washington, DC: Brookings Institution, Center on Urban and Metropolitan Policy, 2000).

25. Rick Layman, "Surge of Population in the Exurbs Continues," *New York Times*, June 21, 2006, National section, http://www.nytimes.com/ (accessed June 23, 3006).

26. David Rusk, *Cities Without Suburbs: Census 2000 Update* (Washington, DC: Woodrow Wilson Center Press, 2003), 17.

27. Eric Damian Kelly and Barbara Becker, *Community Planning: An Introduction to the Comprehensive Plan* (Washington, DC: Island Press, 2000), 260.

28. Owen Gutfreund, *Twentieth-Century Sprawl: Highways and the Reshaping of the American Landscape* (New York: Oxford University Press, 2005).

29. Jackson was quoted in several news articles noting the fiftieth anniversary of the interstate highway system. This quote is from Michael Cabanatuan, "The Interstate Highway System at 50," *San Francisco Chronicle*, June 17, 2006, http://www.sfgate.com/ (accessed June 23, 2006).

30. In a seminal literature review and modeling exercise (applied to California freeways), Robert Cervero supports the argument that highways instigate new development along their paths and induce some traffic that would not have occurred if the highway had not been built; see Robert Cervero, "Road Expansion, Urban Growth, and Induced Traffic: A Path Analysis," *Journal of the American Planning Association* 69 (2003): 145–165. Critics of antisprawl arguments, such as Randal O'Toole and Robert Bruegmann, still doubt the evidence; see Bruegmann, *Sprawl: A Compact History*, 130–131, although Bruegmann seems at least somewhat moved by Cervero's analysis.

31. "Landowners Plan Mega-Mall near DIA," *Denver Post*, January 1, 1999.

32. John Rebchook, "Yellow Brick Road: Development along 470 Beltway Could Reach $35 Billion," *Rocky Mountain News*, January 31, 2003, http://www.rockymountainnews.com (accessed February 3, 2003).

33. Quoted in Rebchook, "Yellow Brick Road." Likewise, the editors of the *Denver Post* saw no irony in reporting that the southwestern arc of the highway was already terribly congested, but that (and this is their good news) the Colorado Department of Transportation had taken over the planning of its last, stalled, leg through the reluctant city of Golden, and that their "renewed hope that the beltway will at last be completed" was because CDOT "is armed with the power to condemn land without the approval of the local government affected." See Editorial, "Metro Highway Loop Coming Around," *Denver Post*, June 27, 2004, http://www.denverpost.com/ (accessed July 4, 2004).

34. Jean Gottmann, *Megalopolis: The Urbanized Northeastern Seaboard of the United States* (New York: Twentieth Century Fund, 1961). Gottmann also identified a midwestern megalopolis anchored by Chicago and a West Coast version extending from San Francisco to San Diego.

35. Robert E. Lang and Dawn Dhavale, *Beyond Megalopolis: Exploring America's New "Megapolitan" Geography*, Census Report Series (Blacksburg: Metropolitan Institute at Virginia Tech, 2005). They even used Google to guide the naming of their megapolitan areas.

36. Gottmann, *Megalopolis*, 5.

37. Besides adding to Lang and Dhavale's list for the West, I would modify some of their assemblages: I would separate out of NorCal a new metrozone in the Great Central Valley ("Great Valley"), and place Sacramento at the articulation of it and the Bay Area's megapolitan swath, which extends to Reno–Carson City. I would take Las Vegas out of Southland because I am impressed with its economic and cultural dynamics, partly driven by gambling, which are independent of the California cities. Finally, I would add St. George, the fastest-growing part of Utah, to the Las Vegas megapolitan area.

38. Mark Sappenfield, "Suburbia's Tide Threatens Identity of Rural America," *Christian Science Monitor*, April 28, 2003, http://www.csmonitor.com (accessed May 1, 2003).

39. The booming towns along the I-25 corridor north of Denver are close enough to several job centers, such as Boulder and Fort Collins, for two-career families to commute in different directions to high-tech jobs. I hear stories from dual-career couples choosing a small town within or just outside the urban corridor and commuting via different angles of attack into the linear jobshed. A real estate broker nicely described this kind of commut-

ing: couples "kiss each other goodbye in the morning, one heads north to Fort Collins, the other south to Longmont or Northglenn and they meet again at night." Quoted in Todd Hartman, "Life on the Edge," *Rocky Mountain News*, June 2, 2001.

40. Utah's fertility rate, twenty-one births per year per one thousand population, is the highest in the United States. Fertility data are from the Centers for Disease Control and Prevention and are published in their National Vital Statistics System (NVSS) reports, available at http://www.cdc.gov/nchs/data/nvsr/nvsr54/nvsr54_02.pdf (accessed April 6, 2006).

41. Interstate highways aided the spread of almost all American cities; early on, these highways acted as intra-urban freeways. Three-term Utah governor Calvin Rampton recalls in his memoirs pressing the Eisenhower administration to allow the states to set interstate construction priorities, and then pressing the Utah Transportation Department to build first the sections of I-15 and I-80 within the Salt Lake City metro area, connecting not states, but suburban cities. Calvin L. Rampton, *As I Recall* (Salt Lake City: University of Utah Press, 1989).

42. At this writing, the highway is under construction, but the path to groundbreaking was littered with objections, especially over wetland loss along the edge of the Great Salt Lake, and fiscal debates. See, for example, http://www.stoplegacyhighway.org/ (accessed April 15, 2006).

43. Morison Institute for Public Policy, *Hits or Misses: Fast Growth in Metropolitan Phoenix* (Phoenix: School of Public Affairs, Arizona State University, 2000), 16–19.

44. Editorial, "Buckeye Awakening to Major Growth That Could Be Bright," *Phoenix Business Journal*, October 31, 2003, http://www.bizjournals.com/phoenix/stories/2003/11/03/editorial2.html (accessed June 30, 2005).

45. Shaun McKinnon, "Growth at the Edges: Development Pushes beyond Valley's Boundaries," *Arizona Republic*, May 12, 2003, http://www.azcentral.com/arizonarepublic/news/articles/ (no longer accessible).

46. Editorial, "Regional Ills, Regional Cures: Inter-City Duels Can't Solve Metropolitan Problems," *Arizona Republic*, March 1, 2001.

47. Chuck Plunkett, "'Super Slab': $2 Billion Road Project Would Use No Public Money," *Denver Post*, March 13, 2005, 1A and 16A.

48. Catherine Reagor, "When Phoenix and Tucson Merge," *Arizona Republic*, April 14, 2006, http://www.azcentral.com/.

49. Quoted in Reagor, "When Phoenix and Tucson Merge."

Chapter 5

1. It has also maintained its pace in the New York region. While the city and close-in suburban counties such as Westchester grew 3–4 percent between 1990 and 2000, the counties farther from the city, but still tied to it, such as Putnam, Orange, and Dutchess, grew 7–13 percent. See David W. Chen, "Outer Suburbs Outpace City in Population Growth," *New York Times*, March 16, 2001.

2. David M. Theobald, "Land-Use Dynamics beyond the American Urban Fringe," *Geographical Review* 91 (2002): 544–564.

David M. Theobald, T. Spies, J. Kline, B. Maxwell, N.T. Hobbs, and V.H. Dale, "Ecological Support for Rural Land-Use Planning," *Ecological Applications* 15: (2005) 1906–1914.

3. Paul C. Sutton, "A Scale-Adjusted Measure of 'Urban Sprawl' Using Nighttime Satellite Imagery," *Remote Sensing of Environment* 86 (2003): 353–369.

4. J. S. Davis, A. C. Nelson, and K. J. Dueker, "The New 'Burbs: The Exurbs and Their Implications for Planning Policy," *Journal of the American Planning Association* 60 (1994): 15–29.

5. A. C. Spectorsky, *The Exurbanites* (Philadelphia: Lippincott Publishers, 1955).

6. John Tarrant, *The End of Exurbia: Who Are All These People and Why Do They Want to Ruin Our Town?* (New York: Stein and Day, 1976).

7. John Herbers, *The New Heartland: America's Flight beyond the Suburbs and How It Is Changing Our Future* (New York: Times Books, 1986).

8. Herbers, *The New Heartland*, 3.

9. Herbers, *The New Heartland*, 4.

10. Herbers, *The New Heartland*, 168.

11. Herbers, *The New Heartland*, 187.

12. Randall Arendt, *Rural by Design* (Chicago: Planners Press, 1994).

13. A. C. Nelson, "Characterizing Exurbia," *Journal of Planning Literature* 6 (1992): 350–368; J. S. Davis et al., "The New 'Burbs"

14. This excellent, and rare, study of large-lot development across a multi-county area was conducted by planning students at the University of Colorado's Senior Environmental Design Studio ("Large-Lot Development in the Denver Metro Region," University of Colorado at Boulder, College of Architecture and Planning). It won the best student project award from the state's chapter of the American Planning Association that year (2000).

15. One researcher, geographer Dave Theobald, has begun pioneering ways to define and map exurbia: see Theobald, "Land-Use Dynamics."

16. Allen Best, "How Dense Can We Be?" *High Country News*, June 13, 2005, 14–15.

17. Tony Davis, "Wildcat Subdivisions Fuel Fight over Sprawl," *High Country News*, April 24, 2000.

18. John Cromartie and John M. Wardwell, "Migrants Settling Far and Wide in the Rural West," *Rural Development Perspectives* 14, no. 2 (1999): 2–8; J. B. Cromartie, "Demographic Trends in the Rural American West" (presentation to the Rocky Mountain Land Use Institute, University of Denver, 1999).

19. William Beyers, "Employment Growth in the Rural West from 1985 to 1995 Outpaced the Nation," *Rural Development Perspectives* 14, no. 2 (1999): 38–43.

20. A Sonoran Institute report concludes that such sources of income are the mainstay of the western economy outside of urban areas: see Ray Rasker, Ben Alexander, Jeff van den Noort, and Rebecca Carter, "Prosperity in the 21st Century West: The Role of Protected Public Lands" (Tucson, AZ: Sonoran Institute, 2004). Walker and Fortmann came to the same conclusion in their detailed study of Nevada County, California: see Peter Walker and Louise Fortmann, "Whose Landscape? A Political Ecology of the 'Exurban' Sierra," *Cultural Geographies* 10 (2003): 469–491.

21. Robert E. Lang, "Office Sprawl: The Evolving Geography of Business," Survey Series (Washington, DC: Brookings Institution, Center on Urban and Metropolitan Policy, 2000).

22. Quoted in Jim Hughes, "'More Not the Merrier' for Some in Park County," *Denver Post*, March 30, 2001.

23. Eugene L. Meyer, "How Far Is Too Far? Developer Plans 4,300 Homes 100 Miles from D.C.," *Washington Post*, August 27, 2005, http://www.washingtonpost.com (accessed September 25, 2005).

24. Laurent Belsie, "Commutes Get Longer and More Rural," *Christian Science Monitor*, May 31, 2002, http://www.csmonitor.com (accessed June 5, 2002).

25. Timothy P. Duane, *Shaping the Sierra: Nature, Culture, and Conflict in the Changing West* (Berkeley: University of California Press, 1998).

26. Ed Marston, "Restoring the West: Goat by Goat," Writers on the Range, *High Country News*, June 24, 2004, available at http://www.hcn.org/servlets/hcn.WOTRArticle?article_id=11305 (accessed July 4, 2006).

27. Susan Ewing, "My Beautiful Ranchette," *High Country News*, May 10, 1999, available at http://www.hcn.org/servlets/hcn.Article?article_id=4994 (accessed July 4, 2006).

28. Auden Schendler, letter to the editor, *High Country News*, July 5, 1999.

29. Walker and Fortmann, "Whose Landscape?"

30. Theobald, "Land-Use Dynamics," 545.

31. W. R. Travis, D. M. Theobald, G. W. Mixon, and T. W. Dickinson, "Western Futures: A Look into the Patterns of Land Use and Future Growth in the American West," Center of the American West Report no. 6 (Boulder, CO: University of Colorado, 2005), available at http://www.centerwest.org/futures.

32. Peter A. Walker, Sarah J. Marvin, and Louise P. Fortmann, "Landscape Changes in Nevada County Reflect Social and Ecological Transitions," *California Agriculture* 57, no. 4 (2003): 115–121.

33. Timothy Duane came to this conclusion by observing landscape patterns in the Sierra Nevada; see Duane, *Shaping the Sierra*.

34. Duane, *Shaping the Sierra*, 204.

35. Felicity Barringer, "Neighbors of Burned Homes Pained by Suburban Sprawl," *New York Times*, December 12, 2004.

36. Duane, *Shaping the Sierra*, 215.

37. As with most exurbs, no formal, inclusive geographic name adheres to this zone, but it encompasses the Conifer-Evergreen area that writer Allen Best used as an exemplar of western exurbia; see Best, "How Dense Can We Be?"

38. Roger Coupol, Gary Beauvais, Dennis Feeney, and Scott Lieske, "The Role and Economic Importance of Private Lands in Providing Habitat for Wyoming's Big Game," Report B-1150, William D. Ruckelshaus Institute of Environment and Natural Resources (Laramie, WY: University of Wyoming, 2004).

39. Southern Rockies Ecosystem Project, *State of the Southern Rockies Ecoregion* (Golden, CO: Colorado Mountain Club Press, 2004), 83.

40. National Park Service, Bandelier National Monument, "Cerro Grande Fire," http://www.nps.gov/cerrogrande/ (accessed November 3, 2006).

41. The problem is large and getting worse, but the government's attempts to define and map the Red Zone have not effectively contributed to an understanding of the real risks of wildfire in the West today; a geographic reality check is in order to counter its overstated assessment. The first major national report was an example of bad geographic analysis, in which the USDA and EPA used vague definitions of terms and indiscriminately lumped whole towns and counties into Red Zones. Large portions of many of the areas they mapped as high-risk areas do not experience the damaging wildfires that blow through forest crowns.

42. Ross E. Milloy, "Population Trends Heighten West's Fire Woes," *New York Times*, August 10, 2000, A-10.

43. David M. Theobald, "Fragmentation by Inholdings and Exurban Development," in *Forest Fragmentation in the Southern Rocky Mountains*, ed. Richard L. Knight, Frederick W. Smith, Steven W. Buskirk, William H. Romme, and William L. Baker (Boulder: University Press of Colorado, 2000), 155–174.

44. Greater Yellowstone Coalition, *Wildfire in the Greater Yellowstone Ecosystem: Toward a Sustainable Future* (Bozeman, MT: Greater Yellowstone Coalition, 2002).

45. Crystal Stanionis and Dennis Glick, "Sprawling into Disaster," in *The Wildfire Reader: A Century of Failed Policy*, ed. George Wuerthner (Washington, DC: Island Press, 2006), 303.

46. David Baron, *The Beast in the Garden: A Modern Parable of Man and Nature* (New York: W. W. Norton, 2004).

47. Tom Knudson, "Majesty and tragedy: The Sierra in peril," *The Sacramento Bee*, a speical five-part report, June 9–13, 1991. This quote is from the June 13 article: "As habitat vanishes, so does wildlife."

48. W. O. Vogel, "Response of Deer to Density and Distribution of Housing in Montana," *Wildlife Society Bulletin* 17, no. 4 (1989): 406–413.

49. Matthew L. Farnsworth, Lisa L. Wolfe, N. Thompson Hobbs, Kenneth P. Burnham, Elizabeth S. Williams, David M. Theobald, Mary M. Conner, and Michael W. Mille, "Human Land Use Influences Chronic Wasting Disease Prevalence in Mule Deer," *Ecological Applications* 15 (2004): 119–136.

50. R. L. Harrison, "A Comparison of Gray Fox Ecology between Residential and Undeveloped Rural Landscapes," *Journal of Wildlife Management* 61 (1994): 112–121; Andrew J. Hansen, Ray Rasker, Bruce Maxwell, Jay J. Rotella, Jerry D. Johnson, Andrea W. Parmenter, Ute Langner, Warren B. Cohen, Rick L. Lawrence, and Matthew P. V. Kraska, "Ecological Causes and Consequences of Gemographic Change in the New West," *BioScience* 52 (2002): 151–162.

51. David M. Theobald, T. Spies, J. Kline, B. Maxwell, N. T. Hobbs, and V. H. Dale, "Ecological Support for Rural Land-Use Planning," *Ecological Applications* 15: (2005): 1906–1914.

52. Readers can see many of Theobald's maps at http://www.nrel.colostate.edu/~davet/. He also developed the model behind the Western Futures maps, available at http://www.centerwest.org/futures.

53. Quoted in Christopher Leonard, "Hot Spots of U.S. Population Growth," *Christian Science Monitor*, June 7, 2005.

Chapter 6

1. Economist Tom Power argues, in his *Lost Landscapes and Failed Economies: The Search for a Value of Place* (Washington, DC: Island Press, 1996), 7–22, that communities hurt themselves by not facing economic reality, especially towns with growing tourist and recreation bases that mistakenly believe their well-being is still tied to the West's extractive economy.

2. Klingenstein's master's thesis on the subject should be required reading for all resort planners: Charles P. Klingenstein, "Issues in Urban Development among Resort Communities in the Intermountain West" (master's thesis, Department of Geography, University of Utah, 1996).

3. Raye C. Ringholz, *Paradise Paved: The Challenge of Growth in the New West* (Salt Lake City: University of Utah Press, 1996), 13.

4. Kurt Repanshek, "A Prophet of Boom: Park City Consultant Myles Rademan Helps Rocky Mountain Ski Towns Grapple with Growth," *Snow Country*, September 1996, 55–57.

5. Much has been written about ski resort development; for critical assessments, see, for example, Hal Clifford, *Downhill Slide: Why the Corporate Ski Industry Is Bad for Skiing, Ski Towns, and the Environment* (San Francisco: Sierra Club Books, 2003); and Hal K. Rothman, *Devil's Bargains: Tourism in the Twentieth-Century American West* (Lawrence: University Press of Kansas, 1998). Rothman wrote about ski, golf, and gateway towns as well as urban resorts such as Las Vegas and Scottsdale. Less work has been done on the dude ranching/hunting/fishing towns, such as Saratoga, Wyoming, that also dot the West.

6. J. Howe, E. McMahon, and L. Propst, *Balancing Nature and Commerce in Gateway Communities* (Washington, DC: Island Press, 1997).

7. P. Messerli, "The Development of Tourism in the Swiss Alps: Economic, Social, and Environmental Effects," *Mountain Research and Development* 7 (1987): 13–24; M. F. Price, "Tourism and Forestry in the Swiss Alps: Parasitism or Symbiosis?" *Mountain Research and Development* 7 (1987): 1–12; H. G. Kariel, "Socio-Cultural Impacts of Tourism in the Austrian Alps, *Mountain Research and Development* 9 (1989): 59–70; C. Pfister and P. Messerli, "Switzerland," in *The Earth as Transformed by Human Action*, ed. B. L. Turner, W. C. Clark, R. W. Kates, J. F. Richards, J. T. Mathews, and W. B. Myers (Cambridge: Cambridge University Press, 1990), 641–652.

8. A. Gill and R. Hartmann, eds., *Mountain Resort Development* (Burnaby, BC: Center for Tourism Policy Research, Simon Fraser University, 1992).

9. David McGinnis, "The Changing Image of Jackson Hole, Wyoming," in Gill and Hartmann, *Mountain Resort Development*, 132.

10. Klingenstein, "Issues in Urban Development," chap. 5, 153–158.

11. R. W. Butler, "The Concept of a Tourist Cycle of Evolution: Implications for Management of Resources," *Canadian Geographer* 14 (1980): 5–12; R. I. Wolfe, "Recreational Travel, the New Migration Revisited," *Ontario Geography* 19 (1983): 103–124; G. Shaw and A. M. Williams, *Critical Issues in Tourism: A Geographical Perspective* (Cambridge, MA: Blackwell, 1994).

12. Shaw and Williams, *Critical Issues in Tourism*, 84.

13. A 2005 community survey in Steamboat Springs, Colorado, found that, given a hypothetical $100 to spend on various community improvements, both second-home owners and full-time residents would spend the single largest amount ($23.22, and $15.77, respectively) on preserving more open space, and less than half that on improvements associated with the ski area. See "2005 City of Steamboat Springs Community Survey," prepared by Linda Venturoni, Northwest Colorado Council of Governments, September 2005, http://www.nwc.cog.co.us/Community%20Surveys/2005%20Steamboat/2005%20Steamboat%20Springs%20without%20interactive.pdf (accessed March 31, 2006).

14. Linda Venturoni, Patrick Long, and Richard Perdue, "The Economic and Social Effects of Second Homes in Four Mountain Resort Counties of Colorado" (paper presented at the annual meeting of the Association of American Geographers, Denver, CO, April 7, 2005), available at http://http://www.nwc.cog.co.us/Second%20Home%20Study/AAGPaperMaster.pdf (accessed March 31, 2006).

15. Power, *Lost Landscapes*.

16. Linda Venturoni, "The Social and Economic Effects of Second Homes" (Silverthorne: Northwest Colorado Council of Governments, 2005), available at http://www.nwc.cog.co.us/Second%20Home%20Study/NWCCOG%202ndHome%20Study%20Binder.pdf (accessed March 31, 2006).

17. Gabe Preston analyzed second homes as part of the Aspen economy in his master's thesis, "Housekeeping in Aspen: Second Homes and Workforce Dynamics" (master's thesis, Department of Geography, University of Colorado at Boulder, 1998).

18. Early ranchers reckoned, and began to cash in on, the tourist value of the Tetons in the late 1880s; see Robert W. Righter, *Crucible for Conservation: The Struggle for Grand Teton National Park* (Boulder: Colorado Associated University Press, 1982), 11.

19. Ringholz, *Paradise Paved*, 13.

20. Town of Jackson, *Jackson / Teton County Comprehensive Plan* (City of Jackson and Teton County Planning Departments, 2002), chap. 1, "Community Vision." See http://www.ci.jackson.wy.us (accessed December 5, 2006).

21. Howe et al., *Balancing Nature and Commerce*. See also R. Rasker, "Rural Development, Conservation, and Public Policy in the Greater Yellowstone Ecosystem," *Society and Natural Resources* 6 (1993): 109–126; G. Rudzitis, "Non-metropolitan Geography: Migration, Sense of Place, and the American West," *Urban Geography* 14 (1993): 574–585.

22. The sight of Boeing 757s and Airbus A310s coming into Eagle, Colorado, or Jackson, Wyoming, always catches me off guard in these unurbanized places.

23. The I-90 improvement project evoked a remarkable land conservation effort, the Mountains to Sound Greenway Trust; see its Web page at http://www.mtsgreenway.org (accessed March 6, 2006). The Colorado Department of Transportation's proposed solutions for weekend gridlock on I-70 between Denver and Vail induced a remarkable new planning entity in the mountain towns and counties—the I-70 Mountain Corridor Coalition—meant mostly to offer alternatives to the department's uninspired, standard proposal: more lanes; see the Coalition's Web page at http://www.nwc.cog.co.us/Rural%20Resort%20Region/I-70%20Coalition/I-70%20Coalition'.htm (accessed April 1, 2006).

24. Edward Stoner, "Vail Valley Sets Real Estate Sales Record: Downvalley Sales Help Drive Up Numbers," *Vail Daily*, January 25, 2006. Similar records were set in Aspen/Pitkin County, other western resorts, and indeed, across the country.

25. Peter Francese, "The Second-Home Boom," *American Demographics/Adage.com*, on-line edition, June 1, 2003, http://www.adage.com.

26. Klingenstein, "Issues in Urban Development," chap. 1, 20–21.

27. Rothman, *Devil's Bargains*; Ringholz, *Paradise Paved*.

28. Park City, "Community Vision '93" (Park City, UT: Planning Department, 1993).

29. The City of Aspen, "2000 Aspen Area Community Plan" (Aspen/Pitkin County Community Development Department, 2000), 26, available at http://www.aspenpitkin.com/pdfs/depts/41/aacp.pdf (accessed September 19, 2005). See also Aspen/Pitkin County Community Development Department, "Aspen Area Community Plan," 1993. The 2000 update of the plan reports that the 60 percent worker housing goal may have been too ambitious, and that they had been unable to achieve it given the community's growing affluence and the associated growing demand for workers; they set, instead, a broad goal of maintaining between 800 and 1,300 affordable units in the planning area.

30. Aspen's "Canary Initiative" includes goals for reducing greenhouse gas emissions and efforts to adapt to global warming; see City of Aspen,

"City of Aspen Canary Initiative," http://www.aspenglobalwarming.com (accessed September 19, 2005). See also the Aspen Global Change Institute's climate impact study *Climate Change and Aspen: An Assessment of Impacts and Potential Responses* (2006), available at http://www.agci.org/index.html (accessed December 5, 2006). I was part of the team working with Aspen on dealing with climate change.

31. Economic Sustainability Committee, "Report and Recommendations" [a]Aspen Chamber Resort Association, 2002), 2–4. See http://www.aspenpitkin.com/depts/41/plan_longrange.cfm (accessed December 4, 2006).

32. City of Aspen, *Aspen Area Community Plan*, February 28, 2000. See especially the updated *Action Plan*, 47–50. Available at http.//www.aspenpitkin.com/depts/41/plan_longrange.cfm.

33. The Jackson Hole Chamber of Commerce has cultivated a program they call "Power of Place" to envision Jackson as a center for global affairs; see Todd Wilkinson, "The Power of Place," *Jackson Hole Magazine*, Summer–Fall, 2005, 42–55.

Chapter 7

1. Fuzzy distinctions—economic, cultural, and geographic—separate the exurban and the rural West. Indeed, geographers and other land use analysts are divided between those who see exurbanization as a fundamentally urban process (as I argue in chap. 5) and those who employ a rural development framework to explain exurbanization.

2. These uses sometimes conflict with each other, as depicted in the film *Shane*: the title character, played by Alan Ladd, protects the homesteaders from the cattle barons who, although they do not own the land, have staked exclusive claim to the "open range." The homesteaders wanted to fence off and cultivate the range; the early ranchers wanted free grass.

3. Peter B. Nelson, "Rural Restructuring in the American West: Land Use, Family, and Class Discourses," *Journal of Rural Studies* 17 (2001): 395–407.

4. J. L. Holechek, "Western Ranching at the Crossroads," *Rangelands* 23, no. 1 (2001): 17–21.

5. The American Farmland Trust has identified ranch subdivision as a major threat to productive ranchlands in the Rocky Mountain states; see American Farmland Trust, *Strategic Ranchland in the Rocky Mountain West: Mapping the Threats to Prime Ranchland in Seven Western States* (Fort Collins, CO: American Farmland Trust, Rocky Mountain Regional Office, n.d.), available at http://www.farmland.org/programs/states/documents/StrategicRanch

landin20thRockyMountainWest.pdf (accessed May 17, 2006). Other studies that describe rural land subdivision in the West include W. Wyckoff and K. Hansen, "Settlement, Livestock Grazing and Environmental Change in Southwest Montana, 1860–1990," *Environmental History Review* 15, no. 4 (1991): 45–71; J. Gersh, "Subdivide and Conquer: Concrete, Condos, and the Second Conquest of the American West," *Amicus Journal*, (Fall 1996): 14–20; W. E. Riebsame, H. Gosnell, and D. M. Theobald, "Land Use and Landscape Change in the U.S. Rocky Mountains I: Theory, Scale and Pattern," *Mountain Research and Development* 16 (1996): 395–405; and Timothy P. Duane's study of the Sierra Nevada foothills, *Shaping the Sierra: Nature, Culture, and Conflict in the Changing West* (Berkeley: University of California Press, 1998).

6. Mark Stevens, "Big Boys Will Be Cowboys," *New York Times Magazine*, November 16, 1995, 72–77, 130–131.

7. Abby Goodnough, "In Florida, a Big Developer Is Counting on Rural Chic," *New York Times,* August 22, 2005.

8. Ingolf Vogeler, *The Myth of the Family Farm: Agribusiness Dominance of U.S. Agriculture* (Boulder, CO: Westview Press, 1981).

9. Pierce Lewis, "The Galactic Metropolis," in *Beyond the Urban Fringe: Land Use Issues in Non-Metropolitan America*, ed. Rutherford H. Platt and George Macinko (Minneapolis: University of Minnesota Press, 1983), 23–49.

10. The Great Plains represent such a broad swath of economic and demographic stagnation that they essentially define the eastern border of what I have called the New West; see William Riebsame, ed., *Atlas of the New West: Portrait of a Changing Region* (New York: W. W. Norton, 1997). Few counties in the plains are growing, and only a handful west of the plains are not growing. A vast literature of "Great Plains studies" documents the economic decline and depopulation of that entire region; see, for example, Donald Worster, *Dust Bowl: The Southern Plains in the 1930s* (New York: Oxford University Press, 1979).

11. Nelson, "Rural Restructuring in the American West," 406.

12. Actually, many of the new homesteaders don't "live" out on the range: they cannot arrange to live full-time in rural landscapes, or may not wish to, yet they buy plots of rural land as second homes, extending a land tenure phenomenon common to the resorts out onto the open range.

13. Paul F. Starrs, *Let the Cowboy Ride: Cattle Ranching in the American West* (Baltimore: Johns Hopkins University Press, 1998).

14. These early amenity ranches are described by Paul Starrs in a detailed study of Sheridan County, Wyoming; see *Let the Cowboy Ride*, 141–158.

15. Physicians, airline pilots, teachers, and professors in particular seem to have a taste for hobby farming and ranching; two of my colleagues, both ecologists, operate significant (160-acre) irrigated farms within 50-minute drives of their offices.

16. Bradley J. Gentner and John A. Tanaka, "Classifying Federal Public Grazing Permittees," *Journal of Range Management* 55 (2002): 2–11.

17. John B. Cromartie and John M. Wardwell, "Migrants Settling Far and Wide in the Rural West," *Rural Development Perspectives* 14, no. 2 (1999): 2–8; Thomas Michael Power, *Lost Landscapes and Failed Economies: The Search for a Value of Place* (Washington, DC: Island Press, 1996); Glenn V. Fuguitt, David L. Brown, and Calvin L. Beale, *The Population of Rural and Small Town America* (New York: Russell Sage, 1989); Nelson, "Rural Restructuring in the American West."

18. Patrick C. Jobes, *Moving Nearer to Heaven: The Illusions and Disillusions of Migrants to Scenic Rural Places* (Westport, CT: Praeger Publishers, 2000).

19. Patricia Nelson Limerick, "The Shadows of Heaven Itself," in Riebsame, *Atlas of the New West*, 161.

20. Interview conducted by Jessica Lage, August 11, 2004, as part of her master's thesis research; see Jessica Lage, "Coming into the Country: New Owners of Ranches in the Sierra Valley, California" (master's thesis, Department of Geography, University of Colorado at Boulder, 2005).

21. Hannah Gosnell, J. H. Haggerty, and W. R. Travis, "Ranchland Ownership Change in the Greater Yellowstone Ecosystem: Implications for Conservation," *Society and Natural Resources* 19 (2006): 743–830.

22. See, for example, Richard Knight, Wendell C. Gilgert, and Ed Marston, eds., *Ranching West of the 100th Meridian: Culture, Ecology, and Economics* (Washington, DC: Island Press, 2002).

23. Julia H. Haggerty, "A Ranchland Genealogy: Land, Livestock and Community in the Upper Yellowstone Valley, 1866–2004" (Ph.D. dissertation, Department of History, University of Colorado at Boulder, 2005).

24. Jon Christensen, "Land Rich, but Cash Poor, in the West," *New York Times*, November 23, 1997, section 3, 1, 11.

25. Haggerty, "A Ranchland Genealogy."

26. Jim Robbins, "Stars Stake a Piece of Big Sky Country," *New York Times*, March 21, 1990, C1, C10.

27. Nelson, "Rural Restructuring in the American West." See also Peter Decker, *Old Fences, New Neighbors* (Tucson: University of Arizona Press, 1998); and D. P. Smith and D. A. Phillips, "Socio-Cultural Representations of Greentrified Pennine Rurality," *Journal of Rural Studies* 17 (2001): 457–469.

28. Census of Agriculture data show that the number of farms and

ranches has actually increased in most western states over the past few decades, while the amount of farmland and ranchland has declined.

29. The Council on Environmental Quality raised the alarm as early as 1976 in a report, "Subdividing Rural America: Impacts of Recreational Lot and Second Home Development" (Washington, DC: U.S. Government Printing Office, 1976).

30. Ranchland subdivision was a major theme of Jeff Gersh and Chelsea Congdon's documentary on western land use, *Subdivide and Conquer: A Modern Western* (Oley, PA: Bullfrog Films, 1999).

31. American Farmland Trust, *Strategic Ranchland in the Rocky Mountain West: Mapping the Threats to Prime Ranchland in Seven Western States* (Washington, DC: American Farmland Trust, 2001). The Ranchland Dynamics project I describe elsewhere in this chapter got its start as a companion effort to this AFT study. Both were funded by The Nature Conservancy and the William and Flora Hewlett Foundation.

32. Starrs, *Let the Cowboy Ride*; Holechek, "Western Ranching at the Crossroads."

33. Riebsame et al., "Land Use and Landscape Change I"; the AFT's study of ranchlands at risk cited ranches in high mountain valleys and around preserves such as Yellowstone National Park and in the southwestern deserts as targets for ownership; see American Farmland Trust, *Strategic Ranchland in the Rocky Mountain West*.

34. See Andrew J. Hansen, Ray Rasker, Bruce Maxwell, Jay J. Rotella, Jerry D. Johnson, Andrea W. Parmenter, Ute Langner, Warren B. Cohen, Rick L. Lawrence, and Matthew P. V. Kraska, "Ecological Causes and Consequences of Demographic Change in the New West," *BioScience* 52 (2002): 151–162.

35. The Nature Conservancy of Arizona, "Ecological Analysis of Conservation Priorities in the Apache Highlands Ecoregion" (Tucson, AZ: The Nature Conservancy, 2003).

36. Riebsame et al., "Land Use and Landscape Change"; D. M. Theobald, H. Gosnell, and W. E. Riebsame, "Land Use and Landscape Change in the U.S. Rocky Mountains II: A Case Study of the East River Valley, Colorado," *Mountain Research and Development* 16 (1996): 407–418.

37. Lage, "Coming into the Country," 78.

38. Mark Haggerty, "Fiscal Impacts of Alternative Development Patterns: Broadwater and Gallatin Counties," *Montana Policy Review* 7, no. 2 (1997): 19–28.

39. Richard Knight, "The Ecology of Ranching," in Knight et al., *Ranching West of the 100th Meridian*, 123–144. See also "Ecological Consequences

When Ranches Die," Knight's testimony to the United States Senate Subcommittee on Public Lands, National Parks, and Forests, September 4, 1992. Results of Knight and colleagues' studies on exurban developments and ranchettes are available in John E. Mitchell, Richard L. Knight, and Richard J. Camp, "Landscape Attributes of Subdivided Ranches," *Rangelands* 24, no. 1 (2002): 3–9; and E. Odell and R. Knight, "Songbird and Medium-Sized Mammal Communities Associated with Exurban Development in Pitkin County, Colorado," *Conservation Biology* 12 (2001): 1143–1150.

40. Knight strongly believes that ranching helps preserve habitat and open space. He makes clear, in some of his articles and presentations, and even in congressional testimony, that he believes that ranching is a culture worthy of preservation, that his views are affected by the friendships he has forged with ranching neighbors at his exurban home in the Front Range foothills, and that environmentalists are wrong to fight livestock grazing because it actually maintains healthy range ecosystems and because, if pressured, ranchers will quit and subdivide their land.

41. Andrew J. Hansen and Jay J. Rotella, "Biophysical Factors, Land Use, and Species Viability in and around Nature Reserves," *Conservation Biology* 16 (2001): 1112–1122.

42. Results of our Ranchland Dynamics research project are available in H. Gosnell and W. R. Travis, "Ranchland Ownership Dynamics in the Rocky Mountain West," *Rangeland Ecology and Management* 58 (2005): 191–198; and Gosnell et al., "Ranchland Ownership Change in the Greater Yellowstone Ecosystem." Unpublished results are available at http://www.centerwest.org/ranchlands.

43. Starrs, *Let the Cowboy Ride*; Haggerty, "A Ranchland Genealogy."

44. Lage, "Coming into the Country"

45. County-level reports from the Ranchland Dynamics assessment of ownership change around Yellowstone National Park are available at http://centerwest.org/ranchlands/ownership.html.

46. Hannah Gosnell, Julia H. Haggerty, and Patrick A. Byorth, "Ranch Ownership Change and New Approaches to Water Resource Management in Southwestern Montana: Implications for Fisheries," *Journal of the American Water Resource Association*, forthcoming.

47. I have visited the ranch twice and observed these practices firsthand, once guided by the owner and once by the ranch manager. On my second visit we were asked to camp in the upper meadows, with the "conservation beef" cattle, to let the wolves know that humans were around. I played loud rock and roll music to announce our presence to the wolves; who knows if they noticed, but my human companions were certainly annoyed.

48. Another aspect of ownership turnover ripe for study is its effect on livestock grazing on federal land. Federal range managers see ranch sales as opportunities to change the use of public grazing allotments—typically to reduce the intensity and duration of livestock grazing—but a comprehensive study has yet to be done.

49. Details of the study of Madison County ranchlands are available in Julia H. Haggerty, Hannah Schneider, Thomas Dickinson, Geneva Williamson Mixon, and William R. Travis, "Ranchland Dynamics in the Greater Yellowstone Ecosystem: Madison County, Montana" (Boulder: Center of the American West, University of Colorado at Boulder, 2003).

50. See George Wuerthner, "Subdivisions versus Agriculture," *Conservation Biology* 8 (1994): 905–908; and Richard L. Knight, George Wallace, and W. E. Riebsame, "Ranching the View," *Conservation Biology* 8 (1995): 459–461. Note here that I joined in this critique of Weurthner's arguments under my previous name, William Riebsame.

Chapter 8

1. Jon Gertner, "Chasing Ground," *New York Times Magazine*, October 16, 2005, available at http://select.nytimes.com/search/restricted/article?res=F00A15FC3D5B0C758DDDA90994DD404482 (accessed October 30, 2005).

2. George Homsy, "Sons of Measure 37: Lessons from Oregon's Property Rights Law," *Planning* 72, no. 6 (2006): 14–19.

3. Timothy P. Duane, *Shaping the Sierra: Nature, Culture, and Conflict in the Changing West* (Berkeley: University of California Press, 1998), xvi.

4. Duane, *Shaping the Sierra*, 350, 351, 360–363, 367, 372–373, 388–390.

5. Oregon's strong statewide planning law was weakened, however, by the passage of Measure 37 on the fall 2004 ballot. California provides significant land use prescriptions through environmental regulations, but even its laudable agricultural land protection system is incentive-based and lacks regulatory teeth, as does Washington's planning and urban growth boundary law.

6. The Colorado Conservation Trust polled local land trusts to build the first statewide database on protected private lands and lands deemed in need of protection; see Colorado Conservation Trust, "Colorado Conservation at a Crossroads: Land Conservation Accomplishments and the Needs, Challenges and Recommendations for the Coming Decade" (Boulder: Colorado Conservation Trust, 2005). At the same time, the Colorado governor called for a statewide compilation and mapping of preserved land, and the state

awarded a two-year contract to create the map (to geographer David Theobald at Colorado State University; see http://www.nrel.colostate.edu/~davet/comap.html). Some conservative voices in the state have said that the map will show that we have preserved too much land: the editors of the *Colorado Springs Gazette* wrote, in a July 7, 2005, editorial, that the map "might convince more people that the relentless, open-ended acquisition by government of private land is a fiscally irresponsible luxury."

7. Former Secretary of the Interior Bruce Babbitt has written an eloquent plea for the vital role of federal land use planning in the social and ecological health of the nation, especially the West: Bruce Babbitt, *Cities in the Wilderness: A New Vision of Land Use in America* (Washington, DC: Island Press, 2005).

8. Ken Snyder, "The Need for Improved Democracy in Planning," Planetizen: The Planning and Development Network, http://www.planetizen.com/node/17469 (accessed October 12, 2005).

9. Ken Snyder, "Putting Democracy Front and Center: Technology for Citizen Participation," *Planning*, June 2006, 24–29.

10. The phrase "informed consent" refers to principles of participatory planning I learned from Hans Bleiker when he and I were professors at the University of Wyoming in the early 1980s. Bleiker has since run a planning consultancy aimed at helping planners and communities get to implementation: http://www.ipm–bleiker.com (accessed January 17, 2007).

11. Eric Damian Kelly and Barbara Becker, *Community Planning: An Introduction to the Comprehensive Plan* (Washington, DC: Island Press, 2000), 19.

12. Kelly and Becker, *Community Planning*, 219.

13. Babbitt, *Cities in the Wilderness*, 5.

14. So, when some city council members and planning staff got serious about addressing the city's jobs-housing imbalance, concluding that they had pretty much done all they could to increase density, they proposed that the city place a limit on commercial space. All hell broke lose. We were laughed at, derided, called unfriendly to business. Some CEOs told the council that if they could not expand they would leave (they could, of course, expand, but only on a flat base of available space). The overwhelming outside response was disbelief: how on earth could any community decide at any point that it had enough jobs? What are they smoking in the People's Republic of Boulder? And, in spite of all this, the city council weighed in to fight the loss of any federal jobs at the National Oceanic and Atmospheric Administration and National Institute of Standards and Technology labs, and the shift of some administrative jobs from the Boulder to the Denver campus of the University of Colorado.

15. See competing op-eds: Vic Smith, "Erie Voters Must Recall Connors, Carter," *Daily Camera* (Boulder, CO), February 1, 2003, 7A; Percy Conarroe, "A Recall? There's No Good Reason for It," *Daily Camera*, February 1, 2003, 7A. The recall effort got the necessary 400 signatures for a vote, but failed at the ballot box.

16. Smith, "Erie Voters Must Recall Connors, Carter."

17. Robyn Morrison, "A Small Town Mayor Challenges Developers: Community Discovers that Once You're on the Growth Train, It's Hard to Get Off," *High Country News*, March 31, 2003, available at http://www.headwatersnews.org/HCN.erie.html (accessed October 17, 2005).

18. Morrison, "A Small Town Mayor Challenges Developers."

19. Editorial, "Choices for Lafayette," *Daily Camera*, October 14, 2005, p. 6B.

20. Editorial, "Choices for Lafayette."

21. Morrison, "A Small Town Mayor Challenges Developers."

22. Mark P. Couch, "E-470 Anchors Land Targeted for Next Suburban Migration," *Denver Post*, July 19, 2001; Jennifer Alsever, "Fort Collins Competes with Loveland for Commercial Growth in Northern Colorado," *Knight Ridder Tribune Business News*, March 3, 2004.

23. The 1962 Federal Highway Aid Act created the "3-C" planning process for urban areas (continuing, cooperative, and comprehensive) and established federal regions that required local involvement in and review of transportation planning. The resulting MPOs emerged to approve, coordinate, and funnel federal money into transportation improvements.

24. Christopher Duerksen and Cara Snyder, *Nature-Friendly Communities: Habitat Protection and Land Use Planning* (Washington, DC: Island Press, 2005), 52.

25. Kelly and Becker, *Community Planning*, 297.

26. American Planning Association, "A Critical Analysis of Planning and Land-Use Laws in Montana" (Chicago, IL: American Planning Association, Research Department, 2001).

27. Babbitt, *Cities in the Wilderness*, 6.

Chapter 9

1. Philip L. Jackson and Robert Kuhlken, *A Rediscovered Frontier: Land Use and Resource Issues in the New West* (Lanham, MD: Rowman & Littlefield, 2006), 174. Jackson and Kuhlken point to Bend as a bellwether of smart growth, but describe a town in transition, still promulgating typical sprawl ("Commercial strip zoning along north-south arterial Highway 97 has created one of the longest four-lane retail promenades this side of Las Vegas"),

but also creating "developments elsewhere in the city [following] the principles of New Urbanism, with high-density residential units built within pedestrian-friendly setting" (174). They conclude, "Only time will tell if enough of these sorts of projects can be stitched together into a richly textured urban fabric that will constitute a cohesive whole. . . . It would be one of the real tragedies in the annals of New West land use if Bend simply plods down the pathway of sprawl, ignoring its own potential to become one of the most livable places in the entire eleven-state region" (175).

2. See also Eric Damian Kelly and Barbara Becker, *Community Planning: An Introduction to the Comprehensive Plan* (Washington, DC: Island Press, 2000).

3. The Headwaters News Web site can be found at http://www.headwatersnews.org.

4. Association of Bay Area Governments, "ABAG Overview: ABAG Programs," http://www.abag.ca.gov/overview/programs.html (accessed November 10, 2006).

5. Association of Bay Area Governments, "Smart Growth Strategy/Livability Footprint Project: Current Status," http://www.abag.ca.gov/planning/smartgrowth/update.html (accessed November 10, 2006).

6. Neal R. Peirce and Curtis W. Johnson, "The Peirce Report 4: No One's in Charge—Time to Replace Our Splintered, Powerless Local Agencies with an Areawide Council That Actually Gets Something Done," *Seattle Times*, October 4, 1989.

7. Eric Sorenson, "Suggestions vs. Action," *Seattle Times*, May 2, 1999.

8. Puget Sound Regional Council, "Vision 2020 + 20: Issues for the Future," conference announcement, http://www.psrc.org/projects/vision/outreach/workshop052005.pdf (accessed November 10, 2006).

9. Unfortunately, few such post-audits have been conducted. For a variety of reasons, both planners and planning agencies, which tend to be prescriptive, not analytical, do little "looking back" at actual outcomes—a process that could guide more effective planning.

10. Joel Garreau labeled the Pacific Northwest "Ecotopia" in his book *The Nine Nations of North America* (Boston: Houghton Mifflin, 1981); he did, barely, include San Francisco in that zone.

11. PLACE^{3}S is described in detail in California Energy Commission, "The Energy Yardstick: Using PLACE^{3}S to Create More Sustainable Communities," http://www.energy.ca.gov/places/ (accessed October 10, 2005).

12. Maricopa Association of Governments, "Valley Vision 2025: Alternatives, Choices, Solutions," http://www.mag.maricopa.gov/archive/vv2025/New%20Pages/vvFrameSet.html; Denver Regional Council of Govern-

ments, "Metro Vision 2030," http://www.drcog.org/index.cfm?page=MetroVision2020.

13. Chris Steins, "Yolo Cities Draw Green Line in Dirt," *Sacramento Bee*, September 5, 2002.

14. Cyndee Fontana, "Two Cities Lay Foundation for Future: Fresno and Clovis Will Concentrate Expansion in the Southeast Area," *Fresno Bee*, August 10, 2001.

15. "Boulder County Countywide Coordinated Comprehensive Development Plan Intergovernmental Agreement," http://www.ci.longmont.co.us/com_dev/pdfs/Boco_IGA.pdf (accessed November 10, 2006).

16. Jon Christensen, "Planning under the Gun: Cleaning Up Lake Tahoe Proves to Be a Dirty Business," *High Country News*, May 12, 1997, 8; available at http://www.hcn.org/servlets/hcn.Article?article_id=3210 (accessed September 1, 2006).

17. Lincoln Institute of Land Policy and the Sonoran Institute, "Growing Smarter at the Edge" (Cambridge, MA: Lincoln Institute of Land Policy, 2005); available at http://www.sonoran.org/programs/growingsmarter.html (accessed November 10, 2006)

18. Christopher J. Duerksen and James Van Hemert, *True West: Authentic Development Patterns for Small Towns and Rural Areas* (Chicago: Planners Press, 2003).

19. See, for example, the Web site of the Community Indicators Consortium, http://www.communityindicators.net (accessed February 10, 2006).

20. Boise's "Ridge to Rivers" open space and trails system is a remarkable, and recent, collaboration of five agencies; see its Web site at http://www.ridgetorivers.org/ (accessed November 10, 2006), and don't miss their panoramic photos from various points on the system!

21. See Joel Berger, "The Last Mile: How to Sustain Long Distance Migrations in Mammals," *Conservation Biology* 18, no. 2 (2004): 320–331, for a call to action and a sober assessment of the task of maintaining migratory pathways.

22. Mountains to Sound Greenway Trust Web page, http://www.mtsgreenway.org (accessed March 22, 2006). The Mountains to Sound Greenway is included as an exemplar in the Conservation Fund's Green Infrastructure Case Study Series; see http://www.greeninfrastructure.net/pdf/mts12.07.05.pdf (accessed March 22, 2006).

23. The plan is detailed in Christopher Duerksen and Cara Snyder, *Nature-Friendly Communities: Habitat Protection and Land Use Planning* (Washington, DC: Island Press, 2005), 253–274.

24. See the Y2Y Initiative's map at http://www.y2y.net/images/map_new_big.jpg (accessed July 21, 2006).

25. The Sonoran Institute spearheaded the state land reform campaign in Arizona and is working to spread the movement across the West. See Sonoran Institute, "State Trust Lands Program," http://www.sonoran.org/programs/si_stl_program_main.html (accessed February 10, 2006).

26. An effort to make comprehensive plans enforceable was under way at this writing in my own state; the Colorado Chapter of the American Planning Association was promoting legislation that would eliminate the "obsolete language" in state planning law stating that a master plan is only advisory. See American Planning Association, Colorado Chapter, "The Colorado Planning Act," http://www.apacolorado.org/legislative_news/2006/Factsheet_HB1053.pdf (accessed February 4, 2006).

27. The Greater Yellowstone Coalition has increased emphasis on community development issues in the ecosystem; see Greater Yellowstone Coalition, "Our Work with the People and Communities of Greater Yellowstone," http://www.greateryellowstone.org/ecosystem/communities/gyc-work.php (accessed November 10, 2006). The Yellowstone to Yukon Conservation Initiative, has paid some attention to community development and sustainability in the zone. Yet Y2Y's atlas of the region neglects land use and settlement as seminal regional features; see Yellowstone to Yukon Conservation Initiative, "A Sense of Place: Issues, Attitudes, and Resources in the Yellowstone to Yukon Ecoregion," http://www.y2y.net/science/conservation/y2yatlas.pdf.

28. See PLAN-Boulder County's Web page at http://bcn.boulder.co.us/planboulder/ (accessed November 10, 2006). The name originally stood for "People's League for Action Now."

29. Such groups include 1000 Friends of New Mexico, http://www.1000friends-nm.org/; 1000 Friends of Washington (now "Futurewise"), http://www.futurewise.org/; and 1000 Friends of Fresno (California), http://www.1000friendsoffresno.org/ (all accessed November 10, 2006).

30. Timothy Beatley, *Native to Nowhere: Sustaining Home and Community in a Global Age* (Washington, DC: Island Press, 2004), 323.

31. William A. Shutkin, *The Land That Could Be: Environmentalism and Democracy in the Twenty-First Century* (Cambridge, MA: MIT Press, 2000), 14.

32. This and other quotes are from Sonoran Institute, "Who We Are: Mission and Vision Statements," http://www.sonoran.org/about_us/si_about_us_mv.html (accessed November 19, 2005).

33. Tom Arrandale, "Rocky Mountain Revamp," *Planning*, January 2004, 10–15.

34. Scot McMillon, "Park County Releases New Growth Policy," *Bozeman Chronicle*, November 18, 2005, http://www.bozemandailychronicle.com/articles/2005/11/18/news/part.prt (accessed November 20, 2005).

35. Sierra Business Council, "Sierra Nevada Wealth Index, 1999–2000 Edition," available at Sierra Business Council, "Publications," http://www.sbcouncil.org/publications.asp (accessed November 20, 2005).

36. Sierra Business Council, "Planning for Prosperity: Building Successful Communities in the Sierra Nevada," available at Sierra Business Council, "Publications," http://www.sbcouncil.org/publications.asp (accessed November 10, 2006). See also Sierra Business Council, "Building Vibrant Sierra Communities: A Commercial and Mixed Use Handbook," available from the same source.

37. The Growth Management Leadership Alliance was created by Henry Richmond, founder of 1000 Friends of Oregon, the much-emulated planning watchdog group formed to monitor and enforce Oregon's land use laws; see http://www.gmla.org/index.html (accessed October 16, 2005). See also the Web site of the Smart Growth Leadership Institute, http://www.sgli.org/index (accessed December 6, 2006).

38. See, for example, Mark Haggerty, "The High Cost of Rural Sprawl," *Greater Yellowstone Report* 13, no. 2 (1996): 1, 4; Dennis Glick, "Necklace or Noose? As Sprawl and Development Encroach on Public lands, GYC's National Conference Explores Solutions," *Greater Yellowstone Report* 17, no. 3 (2000): 4–5.

39. Julia Watkins, "Setting Ecological Guidelines for Rural Residential Development in the Greater Yellowstone Ecosystem" (Bozeman, MT: Greater Yellowstone Coalition, 2005).

40. The Wildlands Project is a large-scale habitat conservation advocacy effort aimed at restoring and reconnecting wildlife habitat across the United States. Its larger-scale conservation plans, centered on maps of habitat cores and corridors that should be protected, regardless of their land use or ownership status, sometimes raise the concerns of local land use interests. See http://www.twp.org/cms/index.cfm (accessed November 20, 2005).

41. Jeff Pidot, "Reinventing Conservation Easements: A Critical Examination and Ideas for Reform," Policy Focus Report 13 (Cambridge, MA: Lincoln Institute of Land Policy, 2005), 1.

42. See the Land Trust Alliance Web site at http://www.lta.org (accessed November 26, 2005).

43. Land Trust Alliance, "Land Trust Standards and Practices: Revised

2004," (Washington, DC: Land Trust Alliance, 2004), available at http://www.lta.org/sp/land_trust_standards_and_practices.pdf (accessed November 25, 2005).

44. See the Montana Land Reliance Web site at http://www.mtlandreliance.org/index.htm (accessed November 27, 2005).

45. See the Web sites of the California Rangeland Trust, http://www.rangelandtrust.org; and the Colorado Cattlemen's Agricultural Land Trust, http://www.ccalt.org/(both accessed November 10, 2006).

46. See the Colorado Conservation Trust's Web site at http://www.coloradoconservationtrust.org (accessed October 10, 2005). Its assessment of conservation needs, "Colorado Conservation at a Crossroads: Land Conservation Accomplishments and the Needs, Challenges and Recommendations for the Coming Decade" (Boulder: Colorado Conservation Trust, 2005), is available at http://www.coloradoconservationtrust.org/goingon/cccexecutivesummary.pdf (accessed November 15, 2005).

47. Bill Birchard, *Nature's Keepers: The Remarkable Story of How the Nature Conservancy Became the Largest Environmental Organization in the World* (San Francisco: Jossey-Bass, 2005).

48. See their current maps at Coalition for Utah's Future, "Envision Utah," http://www.envisionutah.org/ (accessed November 10, 2006).

49. See the Cascade Agenda maps at http://cascadeagenda.com/theREPORT.php (accessed December 6, 2006).

50. SimCity is available from the Maxis Company; see its Web site at http://www.maxis.com.

51. See California Energy Commission, "The Energy Yardstick: Using PLACE^{3}S to Create More Sustainable Communities," http://www.energy.ca.gov/places (accessed November 13, 2005).

52. CommunityViz Web site, http://www.communityviz.com/ (accessed November 10, 2006). Details and case studies of the model's application can be found here. CommunityViz is now marketed through Placeways LLC; see the Placeways Web site at http://shop.placeways.com/index.asp?PageAction=Custom&ID=3 (accessed November 10, 2006).

53. Richard White, *The Organic Machine: The Remaking of the Columbia River* (New York: Hill and Wang, 1996), 64.

54. Shutkin, *The Land That Could Be*, 238.

Index

About Island Press

Island Press is the only nonprofit organization in the United States whose principal purpose is the publication of books on environmental issues and natural resource management. We provide solutions-oriented information to professionals, public officials, business and community leaders, and concerned citizens who are shaping responses to environmental problems.

Since 1984, Island Press has been the leading provider of timely and practical books that take a multidisciplinary approach to critical environmental concerns. Our growing list of titles reflects our commitment to bringing the best of an expanding body of literature to the environmental community throughout North America and the world.

Support for Island Press is provided by the Agua Fund, The Geraldine R. Dodge Foundation, Doris Duke Charitable Foundation, The Ford Foundation, The William and Flora Hewlett Foundation, The Joyce Foundation, Kendeda Sustainability Fund of the Tides Foundation, The Forrest & Frances Lattner Foundation, The Henry Luce Foundation, The John D. and Catherine T. MacArthur Foundation, The Marisla Foundation, The Andrew W. Mellon Foundation, Gordon and Betty Moore Foundation, The Curtis and Edith Munson Foundation, Oak Foundation, The Overbrook Foundation, The David and Lucile Packard Foundation, Wallace Global Fund, The Winslow Foundation, and other generous donors.

The opinions expressed in this book are those of the author(s) and do not necessarily reflect the views of these foundations.

About the Orton Family Foundation

The Orton Family Foundation seeks to transform the land use planning system as a pathway to vibrant and sustainable communities. In partnership with non-profit organizations, local and regional planning agencies and others, the Foundation helps engage and empower people to make land use decisions inspired by their community's heart and soul.

About the Innovation in Place Book Series

The Orton Family Foundation and Island Press have collaborated on a book series that explores the complex land use decisions facing America's communities as they experience rapid changes while striving to maintain a sense of place and personality. The series examines the social, environmental, and economic forces shaping the current land use planning system and presents visions of and solutions for a sustainable future.